Praise for *Your Flight Qu...*

"A perfect companion for the tr...
passengers in today's increasing[...]
travel."
John Wegg, Editor, *Airways* n...

"...extremely readable and interesting. Unlike other books on airline travel, this book relates to passengers with easy, non-technical terms...gives the reader a true appreciation for what goes on 'behind the scenes.'"
Cory Crowell, San Diego Aerospace Museum

"Cronin's book is informative, reassuring, and amusing, and makes for an easy and edifying read while traveling or relaxing at home."
The Atlantic Flyer

"If you are the sort of air traveler who wonders about what's going on both inside and outside of your airliner—both on the ground and in the air—then this book is for you."
Sandy Scherner, The Pilot Shop

"...an insightful and informative look at the commercial airline industry...a fun and easy-to-read book."
Michael Robinson, Jet Up and Go

"Full of useful information for any airline passenger. The section on baggage handling alone is worth the purchase price."
Ross Howell, Jr., Howell Press

"A wealth of information...well worth the purchase price. A great gift."
Walt Bohl, American Aviation Historical Society

Your Flight Questions Answered

by a jetliner pilot

Hub *Airlines run their operations from central "hubs" such as Continental's at Cleveland Hopkins International airport. Continental also has a hub in Newark. For more about* hubs, *see* **Question #17**.
Courtesy of Continental Airlines

Your Flight Questions Answered

by a jetliner pilot

by
John Cronin
Plymouth Press, Ltd

© 1998 John Cronin

ISBN 1-882663-23-3

Plymouth Press, Ltd.
101 Panton Road
Vergennes, VT 05491

On the Web: www.plymouthpress.com

Editing and art direction by Ashley Andersen
Copy-editing by Jan Jones

Plymouth Press has extensive experience in formulating special
editions of our books for educational or promotional purposes.
Regular and special editions of this book are available at a
significant discount when purchased in bulk quantities. To
discuss options or for a free quote, call our marketing manager
at (800) 350-1007.

Printed in the USA

For my parents, John and Marian.
Their support and encouragement
for my dreams knew no bounds.

Contents

List of Questions

35 What are the strips of metal that come up out of the wing during landing?

36 What are the stubby little pieces of wire hanging off the back of the wing?

37 What are the long moving parts on front of the wing?

38 What do the wing flaps do?

39 What does the tail do?

40 What are all those tiny pieces of metal sticking straight up from the wing?

41 What is an aborted takeoff?

42 What is a go-around?

43 What is a missed approach?

44 Why are there different types of airliners?

45 How can I maximize my safety while flying on an airliner?

46 How does a person become a pilot?

47 What sort of training does an airline give a pilot?

48 How do pilots get their work schedule?

49 What do pilots carry in those square-like black cases?

50 How much control do pilots have over departure and arrival times?

51 How does a pilot land in the fog?

Illustrations

Foreword

I decided to write *Your Flight Questions Answered* after a passenger on one of my flights asked me a question…and then another question…and then another question about flying. I realized then that airliner passengers are extremely curious about what they see, hear, and feel while flying, arriving at, and departing from airports.

I've worded my answers to flying travelers' questions so that they can be understood by everyone. *Your Flight Questions Answered* won't make you into a seasoned pilot, but it will give you some basic knowledge of the world of air transport.

Keep in mind that every airliner is different, that every pilot has different techniques, and that every airline and airport have procedures which differ somewhat from the others. Just because something doesn't happen in exactly the way I've explained doesn't mean it's wrong.

As a pilot, I can't resist giving just one bit of general but very important advice: always follow the instructions of the flight crew, as well as those of airline and airport personnel.

I hope you enjoy my book. If you have any comments, questions, or suggestions, please contact Plymouth Press.

John Cronin

Your Questions About Flight Cancellations & Delays

1 Why do flights get cancelled?

Crew problems
A flight may be cancelled if the entire flight crew is not available to fly, for example, if a pilot or flight attendant is sick or injured. The airline then attempts to find a replacement and, depending on the circumstances, the flight may be delayed, not cancelled.

Weather problems
Air traffic control may halt all flights for hours when bad weather hits, whether it affects the airport you're leaving from or your destination airport. Airliners cannot take off unless the visibility at the destination airport is forecast to be at or above a certain distance, usually one-half mile.

Technical difficulties
A flight may be cancelled if an airliner develops a mechanical problem that cannot be quickly repaired, or if a runway is closed. Runways may be unusable for an airliner due to inadequate length, strong crosswind, or other operational restriction.

Avoid the last flight of the day, unless you're prepared to spend the night at or near the airport in case of cancellation. Always try to have an alternate plan if your flight does get cancelled. For example, you might be able to depart from another airport you can reach by car, bus, or train. On one occasion I was the pilot of a flight that passengers had boarded, but air traffic control delayed us for an hour and a half. Then a cargo door broke. It would have taken another two hours to fix, so the flight was cancelled. We had several passengers with international connections who were now stuck because: 1) we were the last flight of the day, 2) there was no other airline flying to the destination, 3) it was too late to drive to another airport to try another flight, and 4) it was the end of the day and the airline had no flights it could divert to pick them up. Had the flight gone as planned, the international passengers would have arrived at the destination with three hours to spare before their connecting flight.

Beware of cancellations of morning flights. In addition to the problems I've already related, there is always the possibility that the airliner designated for your morning flight didn't arrive the previous evening. So don't make the mistake of thinking you can make the first flight, fly uneventfully to your destination, and then rush off to your important business meeting. If your flight is delayed or cancelled, you'll be stuck. This is one situation in which it's definitely better to travel the night before.

In flying, as in life, every once in a while everything seems to go wrong at once. (Pilot joke: "Got time to spare—go by air!")

2 What causes flight delays?

Heavy traffic
The most common cause of delays is traffic volume. For example, if several airlines schedule departures for the same time, a traffic jam is inevitable. On such occasions I've found myself piloting a flight which was eighteenth or further down the line for departure.

Many airports use one runway for departures and another for arrivals. If one runway is closed, then the remaining runway must accommodate both departures and arrivals, and the arriving aircraft must be spaced further apart than usual to allow for departures, which are slotted between arrivals. If arriving airliners get backed up enough, the air traffic control system will not allow any more airliners anywhere to depart for the affected airport. This is one reason why your flight may be held on the ground, even though traffic flow at the airport from which you're departing seems normal.

An air traffic controller is responsible for keeping airliners on *airways* —sort of imaginary traffic lanes in the sky—spaced a certain number of miles apart. If other aircraft already occupy the airway slated for your airliner's flight, then the controller will hold your airliner on the tarmac until a space opens up.

If your flight is unfortunate enough to suffer a long delay, the airline may allow passengers to leave the aircraft. In this situation, I suggest you take your personal belongings with you and do not leave the gate area. Air traffic control may release the flight suddenly, leaving the crew little time to reach the runway for takeoff. The airline then rapidly gathers and re-boards passengers; if you're not immediately available for boarding, you'll miss the flight, compounding your delay. In these situations, pilots try to tell passengers everything they know, which, unfortunately, is often not very much.

Weather

Weather is most commonly the culprit behind delays. When skies are fair, pilots landing airliners indicate to the controller that they have the airport and other aircraft in sight. The controller then clears the flight for a visual approach, leaving the pilot on his own to complete the landing. But when the weather is bad and visibility is poor, the controller has to guide airliners to a final approach course before allowing the pilot to navigate to the runway. In this situation, the controller must allow more spacing than usual between airliners. This slows air traffic considerably.

If thunderstorms are in the landing area, pilots are compelled to fly around them and need extra guidance from controllers. The controllers' workload is increased and flights are slowed. A line of thunderstorms may form what amounts to an impenetrable wall, necessitating that pilots take their airliners into a holding pattern until the storms pass (see *Questions #9* and *#14*).

Storms also wreak havoc with departures. Airliners usually go to one of several *fixes* (a specific navigational location) after departure and then continue on their route. If a thunderstorm is at a fix, air traffic control obviously cannot permit departures to that area. If this happens to your flight, your airliner will pull over or go to a holding area, and you'll watch while other aircraft pass by and depart.

Snow and ice also cause delays. Runways and taxiways may be closed to allow the pavement to be cleared, and airliners must be de-iced, a procedure which takes time, thus causing delays.

My advice is to make every effort to be prepared for disruptions to your schedule. Airliner flights are at the mercy of the air traffic control system, and a problem anywhere along your flight itinerary may adversely affect your chances of reaching your destination on time. And remember, even in this modern era of jet travel, everyone is at the mercy of weather. One bright spot: if your arrival to an intermediate point from which you'll be making a connection is delayed due to weather problems at your destination, there is a chance that your connecting flight will also arrive late. Since late arrivals mean late departures, you might still make your next flight.

Aircraft maintenance

Aircraft maintenance requirements may also delay flights. For example, something as simple as replacing a light bulb may cause a small delay. And if the airliner was late arriving, it takes time to be fueled, cleaned, catered, and loaded with the outbound baggage.

Crew requirements

Another reason for flight delays is that a crew may have *timed out*. Flight crews must have a prescribed amount of rest before and after a workday. The rules are complicated, but when a morning flight is delayed for this reason, it usually means the crew arrived late on the previous evening. The captain of the late flight is responsible for consulting the rule book and ensuring that the crew has at least eight hours of rest. The clock starts 15 minutes after the airliner parks at the gate, and if the morning flight is less than eight hours away, it will be delayed until the eight-hour interval has passed. (Incidentally, the crew has eight hours and 15 minutes to pack up, leave the aircraft, go to the front of the terminal, get a ride to the hotel, check in, get ready for bed, sleep, get showered and dressed, ride back to the airport, and reach the gate. So eight hours of "rest" may mean six hours or less of sleep.)

Keep yourself updated!

If you arrive at the airport and find that your flight is likely to suffer a lengthy delay, don't go off to kill time in a store or restaurant. You should go to the gate and try to find out what's going on. Check the departure monitors at least every fifteen minutes. Here is a story to explain a typical problem:

Your flight is scheduled to depart for Miami from Gate 50. The crew arrives on a Boeing 737 that parks at Gate 10. They are instructed to walk to Gate 50, where another airliner is scheduled to depart with your flight. Meanwhile, the Boeing 737 at Gate 10 is taken, as scheduled, by another crew on another flight.

Now your crew arrives at Gate 50 to discover there is no aircraft—it has been delayed in Chicago because of bad weather there and is now scheduled to arrive one hour after you were scheduled to leave for Miami. The airline adjusts the departure time, adding another 30 minutes for the airliner to be serviced. Total scheduled delay is now 90 minutes.

At this point, everyone is planning on being one and a half hours late. While some passengers disappear to eat, lo-and-behold, the airline discovers they have a spare plane at Gate 25. Suddenly, your flight is scheduled to depart in 30 minutes from gate 25. If you're off shopping or eating, you won't be aware of this new information and you'll likely return to the original gate to discover your flight has departed. The moral here is: if you must leave the boarding area, listen carefully to public address announcements and watch the departure monitors frequently.

❸ My flight was cancelled due to poor visibility, but then the flight crew and airliner took off without me and the other passengers. Why?

The Federal Aviation Regulations were drawn up with one overriding goal: to protect the traveling public. Airliners carrying paying passengers are operating under very strict rules, some of which prescribe how good visibility must be for takeoff. So if visibility is poor, a flight will have to wait for conditions to improve.

However, once a flight has been cancelled, an airliner may operate just like a private aircraft as long as it takes off without passengers, luggage, or other revenue items. Private planes operate under lenient rules compared to those that apply to airlines. For example, they can take off no matter how poor the visibility. For the airliner without passengers, it's up to the crew to decide if it's safe enough to attempt a takeoff. If they think it is, the airline may send the plane and crew somewhere else.

❹ Why did my airliner approach our destination and then divert to another airport?

Unexpected events

There are several possibilities. An airport may suddenly close due to an accident, because of evacuation of the control tower (this happened to me once), or at the time of the arrival or departure of the President of the United States or other VIP.

Weather

A more common reason for diverting is weather. If visibility or weather conditions suddenly and unexpectedly deteriorate so that the pilot cannot, by regulation, attempt an approach, his airliner will be diverted. Failure of a required navigational component during inclement weather, or recognition of danger in the area may cause a diversion. Perhaps airport personnel need to plow snow from the runways and your airliner does not have enough fuel to wait. Or maybe, as once happened to me, runway traction is reported as nil—a pilot can't land if the pavement is a virtual ice-skating rink. Normally, a flight cannot depart unless weather at the destination is forecast to be good enough to attempt an approach (see *Question #1*), but as every passenger knows, sometimes weather forecasts turn out to be wrong.

Mechanical difficulties

Airliners sometimes develop mechanical problems, necessitating diversion of the flight to an airport with a maintenance facility or to an airport with favorable weather or an especially long runway. Other problems may require a landing at the nearest suitable airport.

Section 2

Your Questions About Baggage Handling

5 How is my luggage processed?

When you check your bag at the curb or ticket counter, it's tagged with the final destination ID as well as the flight numbers of all of your flights (see *Figure 1*). At major airports, the bag is then sent to a *sorting area*. All of the luggage checked with each airline which is going to a single destination is placed together on a baggage cart. When departure time draws near, the cart of bags is brought to the airliner and loaded. Depending on the airliner model, the bags may be individually loaded into the baggage compartment or placed in a bin which is then loaded into the aircraft. On the Boeing 727, baggage handlers usually stack the luggage with hard shell exteriors in one section and literally throw the soft luggage and garment bags into another (see *Question #6*).

When you make a connection to another flight, your bags are off-loaded and brought to the sorting area. They then go through the process described above. If your connection is so close time-wise that you've barely made it, there's a good chance your luggage won't have enough time to be processed. If you're lucky, a "runner" will collect your bags and rush them to your connecting flight. But then again it's also possible that your bags will be packed onto the next flight.

Now let's follow a bag which is loaded at Elmira, New York for a trip to Lansing, Michigan, with connections through New York City and Cleveland, Ohio. At Elmira, the bag is tagged with the Lansing identification as well as with all the connecting flight numbers (see *Figure 1*). When the flight arrives in New York City, all the bags are sent to the sorting area. If New York City is a bag's final destination, it is sent up to the appropriate carousel for passenger pickup. Since our particular bag is going on to Lansing by way of Cleveland, it is placed on a baggage cart with other bags going to Cleveland. Our bag is then loaded into a section of the appropriate airliner's cargo hold with other bags connecting through Cleveland, while bags that end in Cleveland are loaded into a different section. Then, when the plane arrives in Cleveland, the baggage handlers take the batch of *through bags* to the sorting area and the other batch directly to the terminal for passenger pickup. Our bag is sorted and placed on a baggage cart for the flight to Lansing. When it arrives at Lansing, it will be offloaded and brought to the terminal to be picked up.

As you can see, your luggage is handled by many people. If just one person misreads the destination tag or the display screen generated by a bar code (see *Figure 1*), you and your bags will end up at different destinations. Another potential problem occurs when the baggage handler reads the tag correctly but places your luggage on the wrong baggage cart. It is important that you remove all tags and stickers from any previous flights. If you allow a ticket agent to do it for you, he or she may miss one, and your bag could end up at last year's vacation spot.

6   How can I minimize the chances of my luggage being lost?

Check in at least one hour before flight time. This will allow you plenty of time for your luggage to be directed to the correct airliner. If you do check in later than an airline's specified preflight time (usually 30 minutes) your bag is considered a *late check*. Late check bags may not have enough time to make it through the sorting area and out to your flight before it leaves, even if you do. At most airlines, if a late check bag does not get to its flight, it is not delivered to your destination address free of charge, unlike its on-time counterparts.

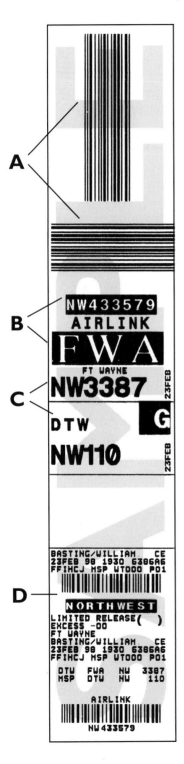

Figure 1. **Baggage ticket**

A. *A set of identical bar codes each containing four characters signify the final destination city.*

B. *Carrier initials* (Northwest) *and a six character bag number are followed by a three-character abbreviation for the final destination* (FWA = Fort Wayne, Indiana).

C. *With the final flight listed first, carrier initials, flight numbers, and date for all legs of the journey make up a mini-itinerary. Destinations for initial flights are listed in smaller type* (DTW=Detroit, Michigan).

D. *The passenger's name and confirmation number (6386AG) are shown. Also here (and on the reverse side of the tag) is a short bar code containing the three character airline code and the six character bag number mentioned above; all flight numbers and destinations; and a passenger locator number* (FFIHCJ).

Sections **A**, **B**, *and* **C** *make up the main portion of the tag affixed to a bag, while* **D**, *and its equivalent on the reverse of the tag, can be removed and placed on airline tickets or collected by agents for service and record-keeping purposes (see* **Question #5***).* Courtesy of Northwest Airlines

Try to get a direct flight. Every time you make a connection, your luggage has one more chance to be misdirected. (That's airline talk for lost.) If you do have a connection, your arrival time and next departure time should be at least an hour apart. That way, if your flight arrives a little late, your luggage will still have enough time to make the next flight, and maybe you can walk, not run, to the gate.

Never pack items of high monetary or sentimental value in luggage that you plan to check. These items belong in your carry-on bag, which should also contain one change of clothes and toiletry items in case your other bags are lost—excuse me, misdirected. And remember to remove all tags and stickers from your luggage after you arrive at your final destination (see *Question #5*).

Never leave your bags unattended at an airport. Keep your leg against them when you're looking around so you're aware if someone tries to move them away. A confused visitor makes an easy target for the skilled thieves that hang around at airports.

I don't recommend putting your home address on any tag that gets attached to your luggage. Anyone reading it will be aware that you're away from home, and may invite themselves over to steal the possessions you haven't packed for your trip. A phone number or the address of your destination on a tag attached to your bags should suffice for directing them to the proper destination in the event they become misrouted.

Finally, buy your luggage to do the job, not because it looks good. I've seen a white suitcase fall off a baggage cart into the snow and disappear far out on the ramp. Bags also fall out of carts at night, so buy bright-colored luggage that stands out and put reflective tape around it. Also, buy good quality, hard shell luggage with a strong handle and good locks. Luggage should always be locked, not only to prevent theft, but also to prevent bags from popping open from the shock of impacts while being handled.

Here's a special tip if you fly on a small airplane or an express airline. There is a slight chance that your bag will not be loaded. Some aircraft, under certain conditions, are not able to take all the luggage and all the passengers because of weight and balance limitations. Because of the high expense of losing luggage, an airline could elect to load all the bags and then remove some passengers, but in reality, it's a lot easier and time-efficient for baggage handlers to leave bags behind. You can minimize your chances of this happening to you by checking your bag at the gate instead of at the main airline ticket counter. When you do this, you're checking a bag that has been through security, and which is then placed on the aircraft immediately. Your bag should be waiting for you at the base of the stairs when you get off the plane. A *gate check* is like a carry-on you put in the baggage compartment. In most cases, this will work only with one or two bags that are the size of carry-on bags—in other words, small.

7   How is my pet handled?

The handling of animals varies somewhat from airline to airline. For starters, dogs for the visually impaired are normally allowed in the cabin, provided they do not obstruct an aisle or emergency exit. Small dogs and cats, transported in a travel cage, can usually be brought on board if they fit under the seat. You should check with your airline about specific in-cabin restrictions for animals.

All other animals are loaded with the luggage. Some airlines may supply a travel cage if you don't have one. Your pet will be handled like luggage, but certainly should not get thrown around or have bags stacked on top of him or her.

The cage normally has a packet attached to it which identifies the animal, owner, destination, and any special care instructions. The pet should be placed in a cargo section of the aircraft which is pressurized and temperature controlled. The cage should also be secured with straps so it doesn't slide around.

The pilot receives a load manifest which indicates the number of adults on board as well as children, infants, cargo or freight, baggage, and live animals. So the captain will know that your pet is on board.

On some airliners, the passengers and baggage are in the same overall compartment, that is to say, the pressure and temperature is the same in both sections. On other airplanes, the baggage compartment may be completely separate and have its own pressurization and temperature control. On at least one occasion I know of, an airliner pilot received an indication of a pressurization failure in a baggage compartment. He knew that a dog loaded there would die of asphyxiation before arrival if the airliner continued on to the planned destination. The pilot diverted the flight and the dog was rescued.

You should consult a veterinarian about air travel for your pet. Medication may be prescribed, depending on the situation. Keep in mind that air travel may be traumatic for a pet.

Section 3

Your Questions About Weather

8 What is weather radar?

Weather radar is a device, like other types of radar, that sends out a signal and processes that signal when and if it returns. If there is nothing in the sky for the radar signal to hit, it is not reflected, and the radar screen appears blank.

The purpose of weather radar is to help pilots avoid storm penetration. A thunderstorm contains large amounts of water in various states—liquid, gas, and solid—and when a radar signal encounters this precipitation, it is reflected back to the aircraft. Based upon the amount of time it takes the signal to return, the weather radar can determine the distance to the storm. The radar can also help determine the severity of a storm by evaluating the intensity of the returning signal. The weaker this signal, the more powerful the storm. Sometimes storms are so strong that most of the signal is returned without having penetrated the entire storm. This is called *attenuation*. Attenuation results in an area within a storm appearing clear on the radar screen. To deal with this possibility, pilots try to get a return signal from the ground surface. If there's nothing there, we have confirmation of severe weather and we know it's time to point the nose the other way.

9 Is lightning dangerous?

Not as dangerous as you might think. Airliners are designed and built so that a bolt of lightning will pass through the airframe harmlessly. It may enter at the nose and exit off of a wingtip or the tail, leaving behind a pinhead-sized mark and scorched area. Rarely, however, it might cause a small hole. If the damage is great enough, the strong airflow generated by the speeding airliner may tear off a piece of the aircraft. This is why lightning strike protocol requires an inspection of the airliner by the maintenance crew after landing. I know of only one instance of lightning possibly being responsible for an in-flight explosion, and that occurred long before the advent of current generations of ultra-safe aircraft.

Actually, the storm producing the lightning is more dangerous than the lightning itself (see *Figure 2*). Such storms contain just about every hazard known to aviation, including: torrential rain, hail, icing, and powerful updrafts and downdrafts as well as rapid and extreme changes in wind speed and direction (See *Question #11*).

Pilots want to see their families again as much as passengers do, so we never intentionally fly into these storms. There are several levels of storm intensity, however, and we may fly near one or weave around several. Passengers may feel as if we are actually flying through storms as we encounter turbulence at the fringes.

Lightning, as well as the storm itself, often seems closer than it actually is. For example, a storm which is over one hundred miles away may appear to be as close as several miles. That's why I use weather radar (see *Question #8*) to tell me the distance of the storm.

10 What is turbulence? Can it be dangerous?

Turbulence which causes a bumpy ride on airliners is associated with three conditions: convection, clouds, and winds. I'll discuss these and then tell you why airliners are safe even in turbulent conditions and how airlines and pilots make sure that you'll be safe when your airliner hits turbulent conditions.

Convection-related turbulence

This type of turbulence is caused by the sun's heating of the ground. Some areas, like parking lots, reflect heat back into the atmosphere, heating the air in that area. As this hot air rises, it cools and eventually sinks, forming ascending and descending columns

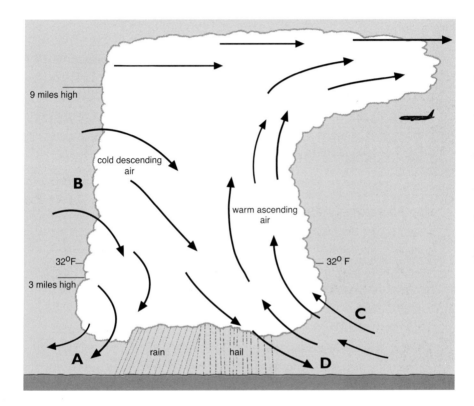

Figure 2. **Thunderstorm** *Thunderstorms can contain almost every major hazard known to aviation, including: torrential rain, hail, icing, and extreme changes in wind speed and direction—* windshear. *Here very cold air from inside the storm cloud descends and spreads out in all directions, while warm rising air is drawn up into the storm. The dangerous result is a surface wind moving away from the storm,* **A** *and* **D***; while wind only several hundred feet higher moves towards the storm,* **B** *and* **C** *(see* **Question #11** *and Figure 3).*

of air. Columns of rising and sinking air caused by convection are most common at lower altitudes. To avoid potentially uncomfortable rides on commuter aircraft during summer months when this phenomenon is worst, consider taking early morning flights.

Cloud-related turbulence

Another form of turbulence occurs in clouds. The most common bumpy cloud is the cumulus cloud—the big fluffy type you see on a summer afternoon. Warm air can hold more moisture than colder air, and as the warm, moist air rises, it cools until the moisture is squeezed out, becoming a visible cloud. The newly cooled air sinks, until it is warmed again and begins another ascent. You have turbulence in this cloud because of the constantly rising and sinking air. So to a pilot, a cumulus cloud is like a signpost that reads "Beware, unstable air ahead."

Wind-related turbulence

Turbulence must also be expected on windy, gusty days. Just as water in a river swirls around obstacles, so does air when it encounters mountains or buildings. When the airflow is disturbed, it moves in different directions at different speeds. As airliners encounter these constantly changing winds, they—and their passengers—experience turbulence.

Effects of turbulence on airliners

Airliners are designed with turbulence in mind: The wings of an aircraft in particular are made to be flexible. Depending on the aircraft, the flexing might be hard to notice—or more obvious than you'd like. Some wings can flex as much as several feet. Because of this, a little turbulence is not a problem, but if it's severe and/or continuous, the pilot will slow the airplane to a *turbulence penetration speed*. This speed, which is designated by the aircraft manufacturer, protects the airliner from too much stress. At or below this speed, the aircraft wing will stall before it breaks. If a column of air rises with enough force, the flow of air over the wing will be disturbed, resulting in a momentary loss of lift, which is far better than breaking the wing. The slower airspeed will also provide a better ride for passengers just as it would if you were driving a car on a bumpy road. If turbulence disturbs you or makes you sick, try to find a seat near the center of the airplane. Your doctor may also be able to prescribe a medication to lessen the reaction to turbulence and prevent airsickness.

11 What is windshear? Is it dangerous?

Windshear is a sudden change in wind direction and/or velocity. Thunderstorm activity is one cause of windshear. Very cold air from a thunderstorm descends rapidly, and when it reaches the ground, it spreads out in all directions. (You may have experienced such a gust at ground level as a storm approached.) While this is happening, the storm is drawing warmer, rising air into itself. This air may be only a few hundred feet above the ground. The result is a surface wind moving away from the storm and a wind aloft moving toward the storm (See *Figure 2*).

Windshear is most dangerous to aircraft during the most critical phases of flight: takeoffs and landings, when the airliner is flying slowly and is close to the ground. Let's take an airliner that must fly at a minimum *airspeed* of about 105 knots (120 mph), or *stall* and lose altitude rapidly. Airspeed is measured relative to the surrounding air, not to the ground, since to maintain lift, an airliner's aerodynamics depend on the rate at which air flows past the wings. Now let's say that the airliner in our example approaches a runway with an airspeed of 122 knots (140 mph). If a 9 knot (10 mph) headwind suddenly shifts into a 9 knot tailwind due to windshear, the airplane's airspeed will immediately drop from 122 knots (140 mph) to 105 knots (120 mph). The airliner, now close or even below the minimum speed required to maintain flight, may drop precipitously and even contact the ground prematurely.

Referring to *Figure 3*, point B is where a pilot landing an airliner would experience an increase in airspeed, thus pushing the plane

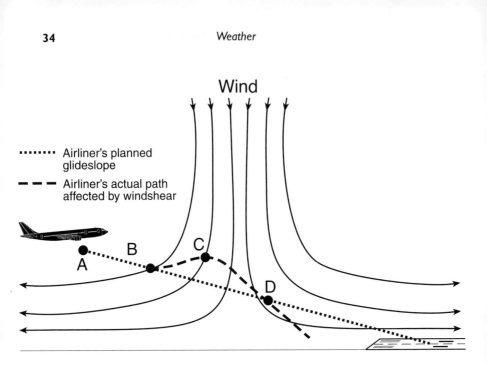

Figure 3. Windshear   *At point **A** the pilot landing an airliner experiences normal flying conditions. At point **B**, as the airliner encounters a headwind, lift increases (because of increased airflow over wing surfaces), thus pushing the plane above the glideslope. At point **C** a pilot may be tempted to decrease thrust to reduce airspeed and return to the glideslope. If he or she does so, between points **C** and **D** the airliner will lose too much altitude, and as the headwind shifts to a tailwind, the aircraft descends rapidly (see **Question #11**).*

above the glideslope (the dotted line representing the planned approach to the runway). Between points B and C, a pilot may be tempted to decrease thrust in order to return to the glideslope. If he or she does so, between points C and D the airliner will experience a sudden loss of airspeed and the plane will descend rapidly. It may be too late for the pilot to increase thrust and the airliner may contact the ground short of the runway.

It is possible for a plane to encounter several shears during the arrival or departure phases of a flight. The experienced pilot will accept a momentary increase in speed when windshear conditions prevail, as he or she is aware that the airliner may gain airspeed only to lose it a few seconds later.

Your Questions About Air Traffic Control

12 **How does air traffic control provide guidance to pilots?**

Air traffic control provides course, weather, and hazard information to pilots, as well as providing the information which allows pilots to keep their airliners at safe distances from one another. Different sets of controllers monitor incoming and outgoing traffic by sight or by radar, throughout an airliner's flight.

The first controller assigned to a departing flight is the *ground controller*. He or she is responsible for all movement at the airport, even automobiles or service trucks that are out on the *ramp* area. This controller tells a pilot what runway to use and the route to take to the runway.

Next is the *tower controller*. He or she is responsible for flights in the immediate airport traffic area, which extends from the surface of the airport to 2,500 feet and out for a five mile radius. This controller has the important responsibility of issuing takeoff and landing clearances.

After takeoff, the pilot contacts the *departure controller*. The flight, now under radar guidance, has its departure directed by a controller who advises the pilot to climb to an initial altitude and to head for the first *navigational fix* of the route. Because this controller is aware of both departing and arriving flights he or she is responsible for guiding airliners safely into and out of airport locales. Departing flights are then directed onto their routes at which time airliners may be as far as 40 miles from the airport.

Finally, the departing pilot communicates with a *center*. Twenty-one centers control all the high altitude airspace in the US. For example, Boston Center, located near Nashua, New Hampshire, controls the airspace from Long Island Sound to Canada, and from the east coast to western New York State. Like all centers, Boston Center is a large windowless building filled with controllers sitting at radar screens.

Once an airliner is on its flight path, the pilot may talk sequentially to several controllers at the same center as the airliner passes through the airspace for which each controller is individually responsible. If the airliner is traveling far enough, the pilot will be handed off to the next center. For example, he or she may talk to Boston Center, then Cleveland Center, and then Chicago Center, until finally arriving at the general destination area.

As the airliner nears its destination, a center controller will instruct the pilot to descend. This phase of the flight customarily begins about 100 to 150 miles from the destination. The center controller then hands the flight over to a *local controller* who passes it to *approach control*.

Approach control routes incoming flights into what amounts to a long convoy. The airliners are spaced so as to allow time for each to approach, land, and taxi clear of the runway. Approach control directs flights onto the final approach course, then hands them off to the tower controller, who issues the final landing clearance. After landing, the pilot is again guided by the ground controller until arriving safely at the gate area.

13 **Why do we taxi onto the runway and then not move for a while?**

Sometimes the air traffic controller will issue an instruction to the pilot to *taxi into position and hold*. The airliner then moves onto the runway and stops with its nose pointed in the direction of departure, ready for takeoff.

This order is most commonly given when an airliner that has just landed is still on the runway from which your departing airliner will take off. It is also given when an airliner is taxiing across your airliner's takeoff runway or when an intersecting runway is in use. Sometimes the order is given to provide extra space between the last departing airliner and your flight. Usually this means that the last departure was made by a very large jetliner, such as a Boeing 747, which generates considerable *wake turbulence*. Allowing this turbulence to dissipate means a safer takeoff and initial flight for your airliner.

14 **Exactly what is a holding pattern?**

A *holding pattern* is a repeating route which an airliner flies over and over until notified by air traffic control to approach or proceed. When there is no wind, it has a oval, racetrack shape. There are speed limits, so pilots generally slow down while making their way to a holding pattern. A standard holding pattern for a jetliner above 14,000 feet would be flown at less than 265 knots (305 mph), would use right turns, and would be run on straight legs lasting 60–90 seconds. Some pilots ask for legs of ten or more miles.

Holding patterns are used when airport operations are unexpectedly suspended. An airport may close to allow for snow removal or in response to an emergency situation, or because the president of the United States is flying in or out of the airport. Airliners also fly holding patterns when there are too many airliners flying into or out of an airport and the traffic system becomes backed up. A pilot might also hold to work on a mechanical problem or to wait for the weather to improve (see *Question #2*).

There may be several aircraft in a holding pattern at the same time, each at a different altitude. Pilots whose airliners are in holding patterns are given a time at which they may expect to be released. Depending on the reason for the hold, the expected duration, and the amount of fuel on board, a pilot may elect to proceed immediately to an alternate airport or choose to hold as long as fuel allows and only then proceed to an alternate destination (see *Question #4*).

15 Are there speed limits for airliners?

Yes. Below 10,000 feet, the limit is 250 knots (288 mph). In an airport traffic area, it's 200 knots (230 mph). There are also limits for holding patterns which depend on altitude and whether or not the airplane is a jet. I typically leave the ground at 144 knots (165 mph), climb out of the area at 170 knots (195 mph), and once above 10,000 feet, accelerate to 325 knots (373 mph). It's easier for the airliner to gain speed with the nose parallel to the ground. If I raise the nose to climb, the airliner slows down, and if I lower it to level off or descend, it speeds up.

16 Who regulates passenger air travel in the United States?

There are currently two major organizations which monitor, and which make recommendations and rules concerning air travel in the United States. One is a government-affiliated agency: the *Federal Aviation Administration*, while the second is the *National Transportation Safety Board*, an independent organization which provides invaluable third-party reporting and recommendations to the air travel industry.

Federal Aviation Administration

The Federal Aviation Administration traces its beginnings back to May 20, 1926, when Congress passed the Air Commerce Act. This legislation established the Aeronautics Branch of the Department of Commerce, which undertook certification of pilots and aircraft and promulgation of safety regulations.

As the air travel industry grew, making constant monitoring of air traffic imperative for safe flights, President Franklin Roosevelt split the existing aeronautics authority into two separate branches: the Civil Aeronautics Administration (CAA) and the Civil Aeronautics Board (CAB). The CAA became responsible for air traffic control, pilot and aircraft certification, safety enforcement, and airway development. The CAB generated safety regulations, provided economic regulation, and performed accident investigations. It was gradually dismantled by the "Sunset Clause" within the Airline Deregulation Act of 1978 and ceased to exist in 1985.

The development of jetliners and the occurrence of several midair collisions during the 1950's pushed Congress into enacting legislation that created the Federal Aviation Agency (later renamed the Federal Aviation Administration) from what had been the CAA. The new agency retained its previous responsibilities but also assumed the former CAB's responsibility for promulgating safety regulations. In the following years, the FAA's responsibilities evolved to include: aviation security, aircraft noise standards, airport aid programs, safety certification of airports, and developing advanced automated air traffic control systems.

Today the FAA also conducts research on advanced air traffic surveillance and control systems, weather radar and communication systems, and efficiency enhancement strategies for high-altitude airliner traffic patterns.

National Transportation Safety Board (NTSB)

On April 1, 1967, Congress established the National Transportation Safety Board to investigate every civil aviation accident and major surface transportation (highway, railway, marine, or pipe-line) accident in the United States, as well as to make recommendations which would serve to prevent future accidents.

Since its inception, the NTSB has severed all connections to the federal government, becoming a fully independent organization which maintains a federal data base on all aviation accidents, provides investigators nationally and internationally for accidents involving aircraft or major aircraft components manufactured in the United States, and acts as an appeals court for any sea or air captain or mechanic who has received disciplinary action from the Federal Aviation Administration.

The NTSB has investigated over 100,000 aviation accidents and thousands of surface transport accidents, becoming the word's leading accident investigation organization. Based upon the findings of these investigations, the Board has issued some 10,000 recommendations regarding all modes of transportation safety. Although the NTSB has no regulatory or enforcement power, its influence over the transportation industry is such that over 80 percent of its recommendations have been adopted by the FAA and other government agencies with enforcement power.

44

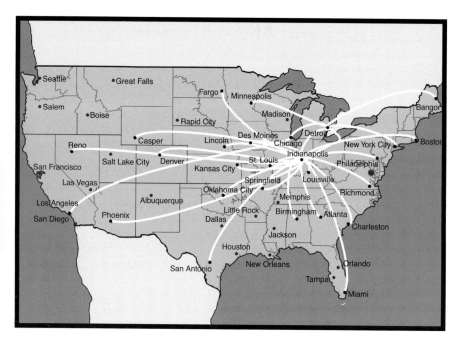

Figure 4. Hub and spoke route system *An airline's* hub *is its regional center of operations. Hubs are connected to other destinations by routes referred to as* spokes. *In the hypothetical example illustrated here, Indianapolis is a primary hub, Detroit serves as a secondary hub, and all other destination cities are* outstations. *Passengers boarding flights at outstations often fly into a hub and board a second flight going to their final destination at another hub or outstation (see* **Question #17***).*

Section 5

Your Questions About Airports

17 **Why do some airports seem to have a lot of airliners from one airline?**

Many airlines use what is known as a *hub and spoke* route system (see *Figure 4*). The airline's regional operation's center–or *hub*–is where all the airliners park at some time during any given day to embark and disembark passengers. If you are making a connection, chances are good that you will be doing so at a hub. Some hubs are: Atlanta (Delta), Dallas-Ft. Worth (American and Southwest), Detroit (Northwest), Newark (Continental), Denver (United), Philadelphia (US Airways), Phoenix (America West), Seattle (Alaska Airlines), and St. Louis (TWA). If you're not at the hub, then you're at an *outstation*. Outstations are connected to the hubs by airliner routes, which can be thought of as *spokes*.

During the night, an airline's aircraft are scattered among the outstations. When morning comes, they take off and fly into the hub, each inbound aircraft carrying passengers who have different ultimate destinations. During the *arrival push*, passengers disembark at the hub, airliners are serviced, and luggage is sorted and reloaded. During the *departure push*, passengers board flights taking a spoke route from the hub to their final destination.

This *in and out* routine occurs several times a day and may lead to delays because of the high concentration of an airline's planes simultaneously moving into and out of the hub. Major airlines have as many as three or four hubs spread across the U.S.

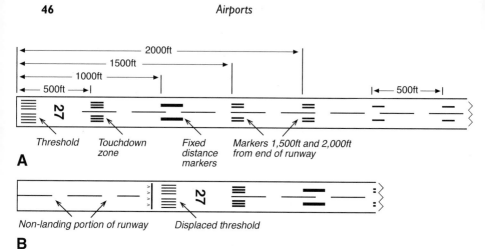

A
Threshold   Touchdown   Fixed   Markers 1,500ft and 2,000ft
             zone   distance   from end of runway
                          markers

B
Non-landing portion of runway   Displaced threshold

Figures 5A and 5B. **Runway markings** **A.** *The* threshold *is the first portion of a runway which may be safely used for landings. The* touchdown zone *provides a maximum safe margin for a landing pilot.* Fixed distance markers *located 1000 feet from the threshold are generally what pilots aim for.* **B.** *A section on either end of a runway may be unusable for landing (due to an obstruction or other problem), in which case the threshold markings are displaced. This area may still be used for takeoffs (see* **Question #18***).*

18 **What do the markings on the runway mean?**

At the beginning of the runway, or *threshold*, are eight white stripes. Next is a number which identifies the particular runway and its approximate direction. When a plane is lined up on Runway 27, its heading should be very close to 270 degrees, or west. Five hundred feet from the threshold is the *touchdown zone*, identified by six white stripes. Five hundred feet farther are two solid white boxes painted on the pavement. These are the *fixed distance markers* identifying the first (or last) thousand feet of runway. At 1,500 feet and 2,000 feet, there are four white stripes and then, continuing at 500 foot intervals, two white stripes, until the markings reverse toward the opposite threshold (see *Figures 5A and 5B*).

The threshold marks the beginning of the usable part of the runway for landing. There may, in some cases, be plenty of pavement before the threshold, but because of an obstruction or other hazard, it's unsuitable for landing. Such a section of runway will have arrows pointing towards the threshold. Although not suitable for landings, this area can still be used to begin a takeoff.

The fixed distance marking is the point the pilot of a landing aircraft aims for. As mentioned above, it's located 1,000 feet from the threshold. Runways which are not equipped with an Instrument Landing System don't have these markings (see *Question #20*).

As your airliner moves from the taxiway onto the runway, it will cross four yellow lines. The first two are solid, while the next two are dashed. The pilot cannot cross the solid lines without the permission of the controller since he or she is taxiing onto a runway. However, the pilot does not need an air traffic controller's permission to exit the runway after landing.

19 What do the letters and numbers on signs near the taxiway mean?

Just like streets in your neighborhood, every *taxiway* has a name. But instead of using a word, taxiways are designated by the letters of the alphabet (see *Figure 6*). Pilots are given taxi diagrams to use at each airport. Believe it or not, taxiing can be more difficult than flying, especially at night or at an unfamiliar airport. The yellow signs help guide the pilot to the runway or terminal under the direction of the ground controller located in the control tower.

Runways are designated by numbers which correspond to their magnetic compass heading. For example, Runway 32 would have a course very close to 320 degrees. Runway signs are conventionally painted red. The letters L, C, or R after the runway number are used to designate parallel runways, for example, 32L (left), 32C (center), and 32R (right).

Figure 6. **Taxiway sign**   *Taxiways are designated by letter-number combinations, and runways by number-letter combinations. The letters L, R, and C indicate parallel runways (see* **Question #19***).* Courtesy of Brian Lassaline/Wayne County's Detroit Metro Airport

20 **What is the little red and white shack I see by the runway?**

In bad weather, which means low visibility, pilots use what is known as the *instrument landing system* (ILS) to navigate their final approach to the runway. The ILS is comprised of two transmitters from which an instrument in the cockpit receives information. One of these transmitters is an antenna next to the red and white shack which provides the pilot with his *glideslope*—or final descent—information (see *Figure 7*). This allows him or her to guide the airliner to the landing end of the runway. The shack houses the electronics for the antenna. The second ILS transmitter/antenna—called a *localizer*—is at the far end of the runway as the airliner makes its approach. The localizer gives pilots the *turn left or right* information they need to stay lined up with the centerline of the runway. This localizer antenna is a contraption of bars resembling a jungle gym at a playground (see *Figure 8*). Not every runway is equipped with an ILS.

While you're looking out of an airliner's window, you might also notice a *transmissometer. This* device is comprised of two platforms. One transmits light; the other receives the light. As visibility deteriorates because of fog, heavy rain, or snow, the amount of light

Figure 7. **ILS antenna**   *Small red and white buildings located near runways house the electronics for the antenna which provides pilots with* glideslope *information needed when using the* instrument landing system *(see* **Questions #20** *and* **#51***)*. Courtesy of Brian Lassaline/Wayne County's Detroit Metro Airport

Figure 8. Localizer antenna *A* localizer antenna *is part of the* instrument landing system. *This antenna provides pilots with left-right information, allowing them to align their airliners with a runway's centerline as they land (see* **Questions #20** and **#51***).* Courtesy of Brian Lassaline/Wayne County's Detroit Metro Airport

received at the second platform naturally decreases. Based on the amount of light registered by the receiver, a measurement known as the *runway visual range* (RVR) is made. When visibility is compromised, the air traffic controller routinely informs the pilot of the RVR, so he or she knows how far down the runway obstructions or other aircraft can be seen. The transmissometer is generally used when visibility is less than 5,000 feet. If the RVR indicates visibility less than 2,400 feet, an airplane can't start the approach unless a long list of requirements is met.

Figure 9. *VHF omni-direction range* *A* VHF omni-direction range (VOR) *emits signals which are unique to each degree of the compass. A cockpit instrument which receives these signals helps pilots to select and maintain their airliner's course (see* **Question #20***)*. Courtesy of Brian Lassaline/Wayne County's Detroit Metro Airport

Some airports also have a navigation station on the field. The most common one is a VOR or VHF omni-direction range. It's white and looks like an upside-down ice cream cone on a saucer with a small shack underneath (see *Figure 9*). This device sends signals to which pilots can tune their omni course indicator (for small planes) or their *horizontal situation indicator* (for large planes). These instruments, in conjunction with the VOR, show pilots whether or not they are following their chosen magnetic compass course (see *Question #51*).

21 **Sometimes, while my airliner is rolling down the runway, I hear and feel a series of thumps. What causes them?**

Many airports, especially large metropolitan airports, have lights mounted in the pavement of the runway called *centerline lights.* They extend from one end of the runway to the other and are particularly useful to pilots when visibility is low. As the plane rolls down the runway, the tires on the nose gear may ride over small metal plates shielding the lights, causing a thump.

22 **What do the various colored lights at the airport mean?**

The following colors are common: green, red, blue, and white. Green lights identify the beginning of the landing portion of the runway, called the threshold. On most runways, this is the beginning of the pavement. Green lights are also placed along the center of taxiways, to aid pilots exiting from runways. Red lights mark the ends of runways and obstructions. Blue lights edge taxiways, and white lights, runways. Some runways also have white lights mounted in the pavement along the centerline.

As an airliner speeds down a runway, the pilot sees white lights on the edges and middle of the runway. When 3,000 feet of runway remain, the centerline lights alternate between red and white. At 2,000 feet, the edge lights are red and at 1,000 feet, all the lights are red.

Many airports have rotating beacons. From the air, these are a green flash followed by a white flash, repeated over and over. Visible from the air long before the rest of the airport, beacons help locate an airport surrounded by the bright lights of a city. Military airports declare themselves with a green flash followed by two white flashes.

Airliners have a red light on their left wingtip, a green light on their right wingtip, and a white light mounted so as to be visible from behind. While in flight, pilots can determine in which direction another airplane is moving based on which lights we see. We also follow the same "rules of the road" as ships. If, for example, an airplane is converging from my right, I will see its red (left wingtip) light and yield. The pilot of that airplane will see my green (right wingtip) light. Airliners also have a flashing red light—which rotates or is a strobe—indicating an operating aircraft and are also equipped with white strobe lights which allow them to stand out against the night sky.

Taxi and *landing lights* are yet another set of white airliner lights, similar to the headlights found on a car. These lights make it easy for other aircraft to see departing and arriving airliners in the congested areas in which departures and arrivals are made.

Taxi lights are normally mounted on the nose gear strut. A variation of the taxi light is called the *trunoff light*, which is used to illuminate the area into which a pilot would like to turn. It shines out diagonally (to the side and forward) allowing the pilot to see the area where the airliner will go. Passengers wont see the taxi light from their seats, but they may notice an illuminated area to the side of the plane when the turnoff lights are selected on. Landing lights are typically mounted along the front of the wing, facing forward and slightly downward so they may illuminate the runway while in the customary nose up landing position (see *Question #37*). These lights are significantly brighter than taxi lights and are mounted either on the fuselage near the wing, or in various spots along the wing, depending upon the particular airliner model. Passengers will notice these lights come on, especially in precipitation.

All lights are turned on for takeoff and while some lights (turnoff and retractable landing lights) are switched off at 10,000 feet, the main landing lights (in front of the wing) may stay on until 18,000 feet, at which altitude they are turned on again in descent.

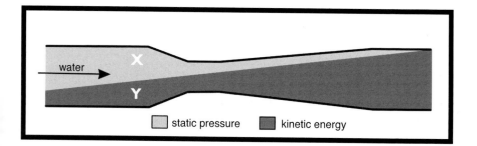

Figure 10. **Venturi tube** Bernoulli's principle, *illustrated by a*
venturi tube, *is a key to understanding what allows aircraft to fly.*
In a venturi tube, as a liquid flows through a constricted area, it
accelerates. As a result, pressure decreases. A similar phenomenon
occurs in the area above an airliner's wing: As air flows over a
wing's upper surface, it accelerates. Pressure decreases here, while
remaining constant under the wing. This creates lift (see
Question #23 *and* Figure 11*).*

Section 6

Your Questions About Airliners

23 **What makes an airliner fly?**

Lift. Many scientific papers and book chapters have been written about it. I will explain the most basic points of the concept—no math formulae or aerodynamic language here. But first I'll need to explain *Bernoulli's principle.*

Daniel Bernoulli (1700-1782) was a Swiss physicist who used a *venturi tube* for demonstrations. A venturi tube—named after the Italian physicist Giovanni Battista Venturi (1746-1822)—is a hollow tube with a narrow throat. As moving fluid encounters the narrow throat, it accelerates in order to continue its forward movement. You can perform this physics experiment at home by putting your thumb over the open end of a garden hose, thus making the water spurt out faster and further.

Here's the technical part. The moving fluid contains X amount of static pressure and Y amount of kinetic energy. For the garden hose analogy, static pressure (X) is the force water exerts against the side of the hose, while kinetic energy (Y) is the energy associated with the actual movement of the water. The sum of X and Y remains constant, because energy can neither be created nor destroyed. If the fluid increases its speed as it passes the venturi throat, thus gaining kinetic energy, it must necessarily have given up some static pressure in order to maintain that constant sum. To put it another way, the fluid has given up pressure in order to accelerate. This is the basis of Bernoulli's principle (see *Figure 10*).

As an airplane accelerates down a runway, Bernoulli's principle kicks in (see *Figure 11*). The top of the wing acts as the narrow part of the venturi tube; air encountering it gives up pressure in order to accelerate, just like in the example of the venturi tube. Air pressure above the wing must thus decrease (remember, the sum of X and Y remains constant), while pressure below the wing

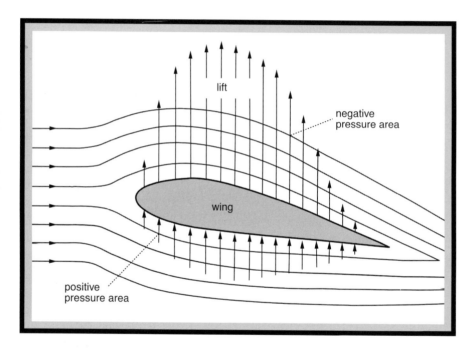

Figure 11. Lift   *As an aircraft accelerates down a runway,* Bernoulli's principle *begins to take effect. The top of a wing acts as the narrow neck of the* venturi tube, *while the underside of the wing acts as the wide mouth of the tube. When air moves across the top wing surface, it gives up pressure in order to accelerate. Meanwhile, pressure under the lower wing surface remains constant. This difference in air pressure near the two wing surfaces produces lift (see* **Question #23** *and* Figure10*).*

remains constant (no Bernoulli effect here). The relatively higher pressure under the wing pushes it up…and lift is created!

24 How does a jet engine work?

There are four words that explain jet engine operation: *suck, squeeze, burn,* and *blow.*

The first step is air intake (suck), followed by compression (the squeeze part). The compressor (see *Figure 12*) has a circular shape with many blades which can be seen by looking at the front of any jet engine. These compressors are driven, through a shaft, by turbines in the exhaust section.

Compressing air makes it very hot, leading to the third step, which is combustion (burn). Fuel is mixed with the hot compressed air and ignited in the combustion chamber. After ignition occurs during engine start-up, burn continues for as long as the engine is in operation.

Combustion further heats the air. In the airliner I fly, the temperature in the combustion chamber during *cruise flight* is about 550 degrees Celsius (over 1,000 degrees Fahrenheit). At low altitudes it can reach 680 degrees Celsius. This hot air exits the combustion chamber with great force, pushing the airliner forward (blow). As the air is exhausted, it turns turbines, which are like windmills. These turbines are connected to the compressors at the front of the engine by a shaft. The turbines' job is to keep the compressors turning, so that more air is sucked in and squeezed.

In summary, you can think of a jet engine as two windmills connected at their centers by a shaft. One windmill, at the rear of the engine, turns the other at the front. As the pilot moves the power levers in the cockpit, he or she is changing the amount of fuel being fed to the fire. This changes the temperature in the combustion chamber and thus the overall *thrust* the engine produces.

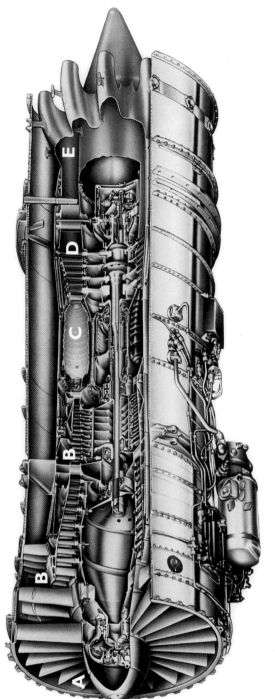

Jet engine *All jet engines perform four basic actions: suck, squeeze, burn, and blow. The parts of an engine which perform these actions are:* **A**, *the air intake, which sucks air in;* **B**, *the compressors, which heat air by squeezing it;* **C**, *the combustion chambers, also called burner cans, where fuel is burned;* **D**, *the turbines which drive compressors and are in turn driven by exhausted—or blown—hot air; and* **E**, *the thrust reverser vanes, which determine which direction the air exits the engine.* Figure of JT8D turbofan jet engine courtesy of Pratt and Whitney

25 What is a turboprop airliner?

A *turboprop airliner* is powered by a jet engine but generates thrust with a propeller.

Basically, one or more turbines—which are like windmills—are placed in the exhaust section of the engine (see *Question #24*). Some turbines are on a shaft that is connected to the propeller, and other turbines are connected to a compressor (which is also found in the pure jet engine). Because the turbines turn rapidly, they are fitted with *reduction gears* so that, for example, on an airliner I flew, a turbine revolving at 19,530 rpm (revolutions per minute) turns a propeller at 1,302 rpm.

This raises the question: Why not just use the jet engine for thrust and forget about adding the propeller? The answer is economics. At the relatively slower speeds and lower altitudes flown by the smaller planes which employ turboprops, a propeller is more efficient than jet exhaust at providing thrust. On the other hand, though, the jet engines available nowadays to turn the turboprops' props are more reliable and easier to maintain than the piston engines that were used to turn props on older airliners such as the DC-3 and Lockheed Constellation.

You may notice a pilot turn the propeller by hand or see it blow around in the wind. This is possible because the propeller is not connected to anything other than reduction gears and the turbines in the exhaust section.

26 Do airliners fly if some parts or systems are broken?

Yes. Airliners have what is called a *Minimum Equipment List* (MEL). If an item does not work, the MEL indicates whether the part or system is required for flight and under what conditions the airplane may fly without it.

If a backup fuel pump does not work, the MEL may say you can fly as long as all other pumps are working and one pump is operated continuously. If the weather radar does not work, the MEL may say you can fly only if thunderstorms can be visually avoided. Some parts, however, must work in order for a flight to proceed: The MEL prohibits flying without them. Depending on the part or system, the MEL may also state a limiting number of days or flights before the part must be repaired or replaced (also known as a *deferred maintenance item*).

27 How is airliner cabin temperature controlled?

The temperature control system varies from airliner to airliner, but the basic operations are the same. Here's how it works:

If you already know how a jet engine operates (see *Question #24*), then you know that the front part of the engine—the compressor—squeezes air to make it hot. Some of the air—called *bleed air*—is directed to the system dedicated to maintaining cabin pressure and temperature, known as *packs* (see *Figure 13*). Once in the packs, some of this hot bleed air continues through plumbing to a mixing chamber, while the rest goes to an *air cycle machine.* In the air cycle machine the hot air drives an expansion turbine. The expansion turbine causes the air to become very cold—the opposite of what a compressor does. This cold air is then sent to meet the hot bleed air in the mixing chamber. A valve regulates the mixing of cold and hot air to obtain the desired temperature. This air mixture is then directed to either floor vents or ceiling vents in the cabin.

28 How are airliners pressurized?

Pressurizing an airliner is at it's most simple level a lot like blowing into a balloon. Here are the details:

As I described in *Question #27,* the engine compressors pump air into the airliner cabin through the packs. A valve in the fuselage

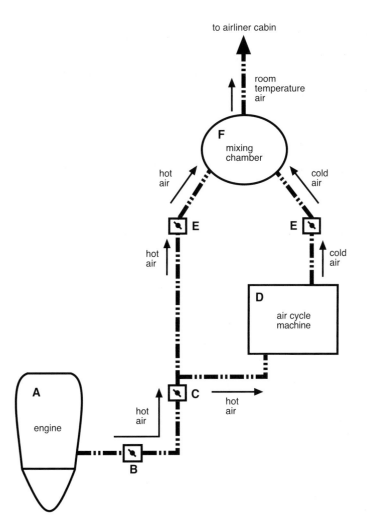

Figure 13. **Packs** *is the term given to the assembly which converts hot* bleed air *into room temperature air directed into the cabin. Air from the engines* (**A**) *enters the system through the* bleed air valve (**B**) *and continues into the* packs *as regulated by the* pack valve (**C**). *Some of the hot air then travels to the* air cycle machine (**D**) *where it is cooled by expansion turbines, while the remaining air travels on toward a* mix valve (**E**). *The newly cooled air from the air cycle machine also travels to a mix valve. These two valves regulate airflow into the* mixing chamber (**F**) *where the combined air is cooled to room temperature before traveling on to the airliner cabin (see* **Question #27***).*

allows a controlled leak to maintain the desired cabin pressure. However, no aircraft is perfectly sealed, and an airliner cabin's pressure cannot be maintained at sea level or whatever air pressure is ambient at the altitude of your departure airport. Aircraft can maintain sea level air pressure in the cabin only up to a certain altitude. In my airliner, we can maintain sea level air pressure up to about 17,000 feet; above this, cabin air pressure must decrease. At 32,000 feet, the pressure in the cabin of my plane is equal to 8,000 feet above sea level.

Before takeoff, pilots set a *pressurization controller* to the cruising altitude they plan to fly at. A chart tells them what air pressure (measured in the altitude equivalent) will be maintained in the cabin at the cruising altitude. A *cabin pressure indicator,* similar to an altimeter, shows actual cabin pressure in terms of altitude in feet and the rate of cabin altitude change. After takeoff, the system allows air to slowly leak out of the airliner until the cabin is at the set air pressure. When initiating a descent, the pilots set the pressurization instrument to the elevation of the destination airport. The system will allow air pressure in the cabin to increase until it's equal to that of your destination. If this isn't done properly (for example, if the system isn't working right), one of two things will happen when you land and open the airliner's door: air will either rush into or out of the cabin, depending upon whether the cabin pressure is higher or lower than the destination airport's air pressure. This problem doesn't often occur on modern jets which rely on automated pressurization systems requiring little pilot input.

29 Sometimes when I'm flying, my ears bother me. What can I do to make them feel better?

The discomfort or pain you may feel in your ears is caused by a difference in air pressure between the inside of your ear and the aircraft cabin. Here's how: As an airliner climbs, the cabin pressure usually decreases to that found at 8,000 feet above sea level. Air which is in the ear got there at the higher pressure present on

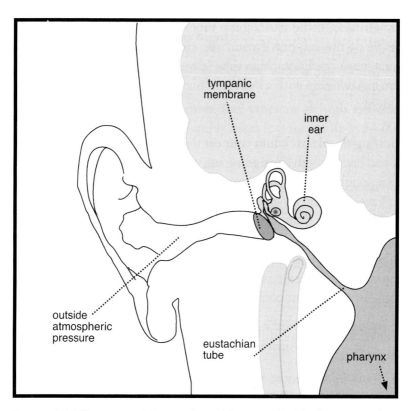

Figure 14. **The ear**   *The eardrum (doctors call this the* tympanic membrane*) forms an airtight seal separating the inner ear form the outside atmosphere. The* Eustachian tube *leads from the inner ear to the mouth (*pharynx*), and normally equalizes pressure between the inside of the ear and the atmosphere. A cold or congestion can interfere with this process, causing pain, or in extreme circumstances, rupture of the eardrum (see* **Question #29***).*

the ground at the departure airport. Now, with reduced cabin pressure, this air pushes against the eardrum (or, the *tympanic membrane*), causing discomfort. Normally, such discomfort is prevented by a very thin tube—called the *Eustachian tube*, shown in *Figure 14*—which connects the inside of the ear to the mouth (*pharynx*), allowing air pressure inside to equalize with pressure outside. The eardrum presents an airtight barrier between the inner and external ear, so no air can escape there. As your airliner reaches cruis-

2 Why do an airliner's engines get quieter right after takeoff?

Once an airliner is safely aloft, the power from the engines is reduced to an initial climb power setting. Not only is this good for the engines, but it reduces noise to the communities below.

The amount of power required for takeoff is computed before the flight. If a plane requires less than 100% power (thrust), then the pilots may use the reduced power setting and save a little wear and tear on the engines—not to mention the ears. In an emergency, a pilot can always push the power levers as far as they will go. The engines might need to be replaced afterwards, but they are capable of producing enormous thrust for a short period of time if necessary.

3 As the airliner I'm flying in prepares to land, the sound of the engines constantly changes. Why?

The short answer is that getting ready to land requires frequent changes in engine thrust. Here's why:

As an airliner approaches an airport, the air traffic controller often restricts it to a particular airspeed. Just like the driver of a car, the pilot must then reduce the thrust of the airliner's engines. Pilots know that a particular power setting results in a particular airspeed and thus reduce power as required. If the controller has given instructions to slow immediately, then power may be suddenly and substantially reduced in order to expedite the speed reduction. Then, as the airliner reaches the desired airspeed, power is increased to maintain it.

If the airliner is cleared to a lower altitude, the power must be reduced again. Again, if you think about the situation in a car, if you are cruising at 55 mph and then start going downhill, your speed increases unless you reduce power. The same is true for an airliner: every time the pilot deceases altitude, he or she needs to reduce the power setting to maintain the assigned airspeed.

Changes in wind direction and velocity at different altitudes may also affect airspeed, so a pilot may frequently adjust—*milk*—the power levers to maintain a particular airspeed and descent rate during the approach to a landing (see *Question #11* and *Figure 3*).

34 **Why do the airliner's engines suddenly get loud right after the airliner touches down on the runway?**

Upon landing, the pilot usually needs to apply *reverse thrust* to slow the airliner. Reverse thrust is most effective while the plane is going fast; as the airplane slows, reverse thrust is decreased or discontinued and tire braking like an automobile's is applied.

On a jet engine, reverse thrust is applied by redirecting the exhaust gases forward. On some Boeing 737s, you may see the sides of the engine open up. The ends of both sides then meet to block the exhaust and direct it forward. On other airplanes, you will notice the rear half of the engine (actually an outer casing) move rearward. Again, the normal exhaust is being blocked and redirected forward. At the same time that thrust is being redirected, it is increased to maximize braking. This increase in thrust, along with the forward deflection of exhaust, results in a sudden increase in the noise level inside the cabin.

In the case of a propeller-driven airliner, the blade angle of the propeller is changed to a setting which pushes air forward rather than rearward, and, as with the jet, thrust is increased. However, not all propeller airplanes are capable of reverse thrust.

35 **What are the strips of metal that come up out of the wing during landing?**

These are known as *speed brakes* or *spoilers* (see *Figure 15*). When the plane lands, the pilot wants to get the airplane's weight down

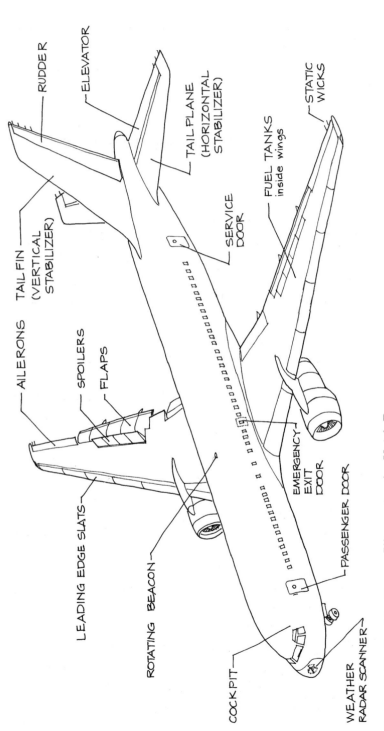

Figure 15. *Typical jetliner* Illustration by Keith Brown.

onto the landing gear, which allows for better braking and control. By raising spoilers, the pilot disrupts (or spoils) the flow of air over the wing, destroying much of the lift. The spoilers also create drag, which helps to slow the airliner. You may sometimes observe these devices being deployed during a flight. Don't worry; it just means the pilot is going to descend very rapidly. Without getting into aerodynamics, I will give you a basic idea of how they work.

Let's say the airplane is moving at an airspeed of 250 knots (218 mph) and descending at a rate of 2,000 feet per minute. The air traffic controller may for some reason need the pilot to expedite the descent. When the pilot extends the spoilers, he or she creates drag and disrupts lift. The pilot then can either maintain the descent rate and reduce airspeed, or maintain the airspeed and descend at a faster rate.

Sometimes a plane may be reduced to a slow airspeed and then be instructed to descend. If the pilot finds that descending will result in an increase in airspeed, he or she can use the spoilers to maintain the assigned airspeed while still performing a descent.

36 What are the stubby little pieces of wire hanging off the back of the wing?

You may observe up to five short, stubby wires protruding from the back of a wingtip or from the tail. These are called *static wicks* (see *Figure 15*).

What are they for? An airliner flying through electrically charged air near a storm, for example, develops a static charge. The static wicks are a means by which this accumulated static electricity can be discharged. St. Elmo's fire is the name given to the luminous glow sometimes visible when there is a large static discharge. On a pilot's windshield, it may appear as webs of electricity, dancing and flickering like little lightning bolts.

37   **What are the long moving parts on front of the wing?**

These parts are called *leading edge flaps* and *leading edge slats*. Located on the front of the wing (see *Figure 15*), they—like flaps (see *Question #38*)—change the shape of the wing in order to maximize lift in some specific situations. What are these situations? Allow me to explain…

Airliner wings are designed to provide lift ideally when the airliner is travelling forward with the fuselage parallel to the ground. Unfortunately, this is not always the case, for example, when taking off or landing. The crucial relationship of the wings to the airstream is called the *angle of attack*. Watch a jet coming in to land or taking off. The airliner will be pointed towards the sky in both these situations. As a result of the high angle of attack, air is coming at the wings from below and not head-on. Since, as I mentioned above, the wing is designed to provide lift when the wing's angle of attack is head-on, lift is partially diminished in this situation.

Leading edge slats and *flaps* are designed to optimize airflow over the upper surfaces of the wings when the angle of attack compromises lift. A leading edge slat maximizes lift during a high angle of attack by acting as a scoop. The slat folds down from the front of the wing and a large curved area is formed. When deployed in this manner, the scoop directs air to the top of the wing, instead of letting it slide underneath. Leading edge flaps extend outwards from underneath the wing, meeting the slat to a make a more effective scoop. They are usually located on the part of the wing closest to the fuselage, and pivot down from there so that, when fully deployed, they appear to line up with the slats.

Deploying something called a *slot*, on the other hand, creates an opening between the front and the body of the wing (see *Figures 16A* and *16B*). As a result, air flows up from under the wing, is guided through the opening, and then continues smoothly over

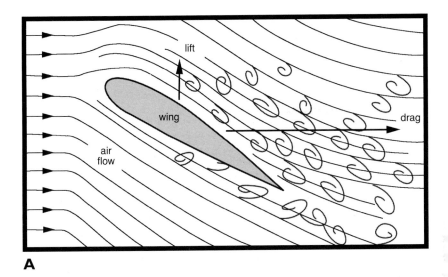

A

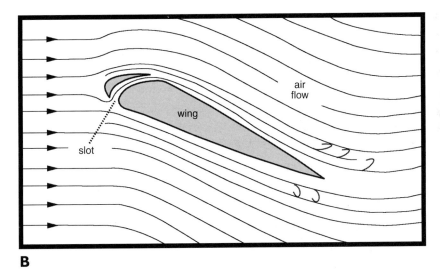

B

Figures 16A and 16B. ***Wing slot*** **A.** *When flying with a high* angle of attack, *airflow over the top surface of an airliner's wings is disrupted, creating drag and diminishing lift.* **B.** *To maintain lift during landing and takeoff, some airliners are equipped with* slots. *Slots create an opening in the front edge of a wing which allows air to flow from under the wing up and over the top of the wing surface. This sustains lift (see* **Question #37***).*

the upper wing surfaces, maintaining lift in a situation where it would otherwise be compromised.

38 What do the wing flaps do?

Wing flaps, at the rear edge of the wing, change the shape of the wing when they are extended and also increase the total surface area of the wing (see *Figure 15*). This results in increased lift, which allows the pilot to maintain flight at low airspeeds. This is important during landing because it enables the pilot to slow down prior to touching the airliner down on the runway. Without flaps, an approach would have to be made at a much higher speed, necessitating a considerably longer runway. Deploying flaps allows a pilot to significantly slow the airliner before touching down on a runway. During takeoff, on the other hand, flaps increase lift, so that the airliner can depart using a much smaller length of runway.

When the flaps are lowered, as for landing, they also increase drag (which is why they're not deployed during most of the flight). The airplane slows down and at some point the pilot will have to increase power to overcome the drag and maintain airspeed. This is why you may notice a change in the sound of the engines shortly after the flaps are adjusted, especially during the approach to landing (see *Question #33*).

39 What does the tail do?

The tail is used to balance an aircraft.

An airliner is like a seesaw that pivots on its wings. Since the front end is pulled down by gravity, something is needed to push the rear end down to maintain balance. The *little wing,* or *tailplane,* is also referred to as a *horizontal stabilizer.* Shaped like an upside-down wing, it generates lift on the bottom—actually, reverse lift— causing the rear end of the airliner to be pushed down.

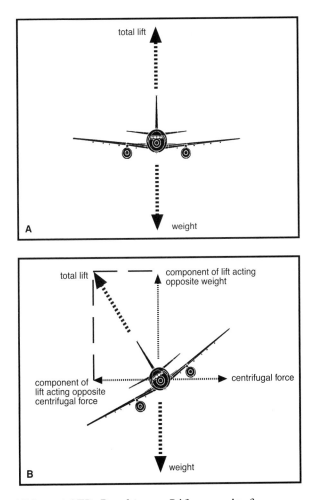

Figures 17A and 17B. **Banking**   *Lift exerts its force perpendicularly to an aircraft's wing and to the direction in which the aircraft is moving; it also acts in the direction opposite to an airliner's weight. In* Figure 17A, *lift acts straight upwards because the aircraft's wings are completely horizontal. In* Figure 17B, *when an airliner turns or* banks, *centrifugal force pulls the craft toward the outside of the turn—just as in a car—but one component of lift opposes this effect, while the other continues to act in the direction opposite to the airliner's weight. The resulting total lift acts in a direction precisely between these two components, turning the airliner until the pilot returns its wings to a horizontal position (see* Question #39).

Mounted onto the horizontal stabilizer is the *elevator* (see *Figure 13*). The elevator is manipulated by the pilot pulling or pushing on the *control wheel* in the cockpit. It controls the airplane's *pitch*, which is the up and down attitude of the nose.

The part of the tail sticking up is the *vertical stabilizer* or *tailfin*. The rudder is attached to it and is used to turn an aircraft (see *Question #40* and *Figure 15*). By pushing the rudder pedals at his feet, the pilot moves the rudder left or right, resulting in, respectively, the nose being pushed to the left or right.

An aircraft turns when the pilot *banks* it, that is, when an aircraft rotates down towards one wing or the other (see *Figures 17A* and *17B*). In level flight, the lift from the wings pushes the plane up. If the wings are banked, or on an angle, the lift is also being banked and pushes the plane into a turn. The pilot has to use the rudder so the tail follows the plane through the turn smoothly. If, in a left turn, the pilot does not use enough rudder, you feel yourself being pulled to your left. This is called a *slip*. If, on the other hand, the pilot uses too much rudder, you feel yourself being pushed to the right. This is called a *skid*. When just enough rudder is applied, you feel like gravity is holding you naturally in your seat, no matter how steeply the plane banks.

40 **What are all those tiny pieces of metal sticking straight up from the wing?**

They are commonly referred to as *vortex generators* (see *Figure 18*). You may see vortex generators on top of the wing or on the side of the tail. As the name implies, they create a vortex, which prevents air from separating from the wing or other surface area on an aircraft. This aids in maintaining the effectiveness of controls.

By controls I mean the *ailerons* (tiny flap-like devices near the wingtips) and the *rudder* (the moving part of the tail which travels to the left or right) (see *Question #39* and *Figure 15*). Vortex generators are most commonly located just in front of the aile-

Figure 18. **Vortex generators**   *A* vortex generator *is a small piece of protruding metal attached to either an aircraft's wing or tail. It is designed to maintain effectiveness of controls such as the ailerons, flaps and rudder. The vortex generators each create a vortex—a spiralling mass of air that sucks everything nearby into its center. The vortices pull air closer to the control devices located on an aircraft's wings and tail. (see* **Questions #38 and #40***).*

rons. You might get a glimpse of a vortex trailing off a wingtip during damp or misty days; it looks like a sideways tornado.

41 What is an aborted takeoff?

An *aborted takeoff* is a termination of a takeoff by the pilot at any time after takeoff clearance has been received from the control tower.

During takeoff, one pilot is responsible for steering the plane down the runway. The other pilot monitors the engine instruments to make sure the engines are operating normally. If any indication is outside of acceptable limits or just not normal, the takeoff may be aborted. Pilots are trained so that the illumination of any warning light or activation of any oral warning system will trigger an immediate abort. The philosophy here is that there is no time to look into what is wrong—we should just abort and then figure out what's going on.

An aborted takeoff can be risky. Before the flight, pilots figure out an airspeed known as V1. This is the *go/no go speed*. Once this airspeed is reached during takeoff, the airliner is committed to flight. If an airliner is just about to reach V_1 and an abort is initiated, the pilot will probably apply maximum braking. Every situation is different, but sometimes the use of maximum brakes can cause the tires of the landing gear to blow out or even ignite due to the heat built up by friction.

Normally, aborted takeoffs aren't so dramatic. You feel the airliner accelerating and then, when the abort is initiated, the speed brakes come up out of the wing, the engines go into reverse (which is noisy—see *Question #34)*, and the pilot applies brakes to the tires. It's no big deal except for the paperwork. That's another story.

42 What is a go-around?

A *go-around* is an aborted approach to landing executed when it is unsafe to continue the approach or to land. It may be necessary when the pilot determines that the aircraft is not properly configured to land (perhaps the landing gear didn't fully extend), is on an improper approach path, or encounters a severe windshear or wake turbulence from the last airliner following the same approach.

A go-around may be initiated by either the air traffic controller or pilot. For example, if the controller sees that the preceding landing aircraft cannot get off the runway fast enough, he or she will issue the go-around instruction.

A go-around is initiated by applying a predetermined amount of power. Based on air temperature and field elevation, the pilot might apply from 90% to 100% power. The flaps are usually brought to a takeoff setting and the landing gear retracted. Every situation and airplane is different, so the procedure varies. Once the airliner is climbing out of the area, the pilot will be placed back in the approach sequence or, depending on the situation, diverted to another airport or to a holding pattern (see *Questions #4* and *#14*).

43 What is a missed approach?

The flying of the airliner in this circumstance is similar to that of a *go-around* (see *Question #42*), but usually occurs as a result of bad weather. During a *missed approach*, a pilot flies a specific course to land at an airport and, at a pre-designated point, fails to see the runway and thus cannot land.

The *instrument landing system* (discussed in *Questions #20* and *#51*) guides the pilot up and down, and left and right, all the way to the appropriate landing point on the runway. For many runways and airliners, the pilot can descend to a point 200 feet above the ground using this system—that's pretty low and just a few seconds from landing. If, at this point, nothing identifiable as part of the runway is in sight, the pilot must execute a missed approach. If the pilot does see something related to the runway, he or she can descend to 100 feet above the touchdown zone elevation. The Federal Aviation regulations are very specific about the details.

Once the approach has been missed, the aircraft is flown in the same manner as during a go-around and the pilot must comply with the missed approach procedure for that particular approach. This prescribes what altitude to climb to, the heading to fly, and where to navigate. Many pilots might opt to give the approach a second try before going to another airport, but this decision depends on a number of factors.

44 Why are there different types of airliners?

There are two basic reasons for this: airliner job descriptions and competition. Let me elaborate:

First, different jets are built to do different jobs. For example, there is the *short-haul* market, which requires frequent flights of short duration. An airline might run a flight between New York and Boston every hour like a bus. Because the flights are so frequent, there is not much time for passengers to accumulate. As a result, flights frequently depart with fewer than 200 passengers. For this job, a small jet, like one of the Boeing 737 series, is appropriate.

At the other end of the spectrum is the *long-haul* world of intercontinental travel. Flights over oceans take many hours and require a lot of fuel. An airline might operate only two or three flights a day, so a large number of passengers may accumulate. For this job a large jetliner, like the Boeing 747, which can carry over 500 passengers, fits the bill.

The second reason for a diversity of airliner models is competition between manufacturers. For example, for short-haul flights, Airbus produces the A320 to compete with Boeing's 737; and for long-haul flights Airbus produces the A330 and A340 to compete with Boeing's 747, 767 and 777. An airline interested in buying airliners examines not only price, but how much money per *seat mile* (one seat occupied by one passenger flying one mile) they will cost to operate, their mechanical reliability, and the ease and cost of carrying out maintenance. Availability of parts and manufacturer support are also important considerations.

45 How can I maximize my safety while flying on an airliner?

Pay attention to the flight attendant's (or pilot's) safety briefing or—on some airliners—the video presentation. Review the safety card, usually located in the pocket in back of the seat in front of you. If you don't understand something, ask the flight attendant. During an emergency, follow the directions of the crew—they are highly trained to deal with special situations.

Many airliners today have aisle path lighting. These lights lead to exits. They are activated either manually by the crew or automatically by the electrical system. But as a backup to these lights, you should know how many rows to crawl to reach the exits behind you and in front of you.

Avoid certain clothes, such as pantyhose (leg burns if ignited), neckties (strangulation risk) and skirts (long skirts can hamper your evacuation). Avoid the use of hair spray (it's flammable).

Sitting next to an emergency exit is not a responsibility to be taken lightly. People's lives may depend on you. You should be thoroughly knowledgeable in the operation of the exit and when it should be used. On some aircraft, for example, certain exits should not be opened after a water landing; you could flood the cabin. The safety briefing card explains this. By the way, if you like a lot of leg room, you will enjoy the exit row seats. They are usually wider to allow people to exit (be advised that you may not be able to recline your seat). If you are uncomfortable with the extra responsibility, ask to be seated somewhere else.

If you sit in an aisle seat, be on guard every time the overhead bin is opened. If you're not alert, a fellow passenger may open the bin and a briefcase or bag could fall on your head.

Wear comfortable shoes with laces or some sort of fastening device. If you wear sandals or high heels, you will probably lose them during impact or evacuation. You then have to go barefoot through the wreckage which will be littered with hot, twisted and sharp pieces of metal, plastic, and glass. Wear pants. They provide better protection than shorts or a skirt. I cannot tell you the tail section is safer than the front or an aisle seat is better than a window seat. Every accident and airplane presents a unique situation. Fortunately, accidents are very rare.

If you're going scuba diving away from home, wait for 12 hours after your last dive before flying. Airliner cabin atmospheres are at considerably lower pressure than that at sea level, which could put you at risk for getting decompression sickness—known as "the bends." And if you undergo extensive dental work, it may also be wise to wait several days before flying—air bubbles which get lodged in tissue may expand as the airliner gains altitude, causing considerable pain.

On a final note, you should be sober and drug free—you will have a tough time evacuating an aircraft if you are under the influence of alcohol or drugs. In case you didn't know, an airline cannot board a passenger who appears to be incapacitated (drunk or high).

Your Questions About Pilots

46 **How does a person become a pilot?**

There are two routes: military and civilian.

First, I'll explain the military route. After volunteering for the military, a person must be accepted into an officer training program. Following successful completion of this program, an individual must be qualified and selected to attend pilot training. After graduating from pilot training, which usually takes one to two years, he or she will get wings. The fledgling military pilot then goes on to instruction on flying a specific assigned aircraft, as well as undergoing survival training.

If a retired military pilot wants to become a civilian pilot, and has flown actively within the past year, he or she brings his or her flight records into the Flight Standards District Office and is issued the equivalent of a civilian license. If the pilot has not flown in the past year, he or she must take and pass flight tests with a representative of the FAA (see *Questions #16* and *#47*).

The civilian route involves obtaining one or more (or all if you plan to become a professional pilot) of the following licenses, which grant increasing flight privileges and responsibilities:
- *private*
- *commercial*
- *instrument*
- *flight instructor*
- *multi-engine*

- *instrument instructor*
- *airline transport pilot*

Private pilot's license

The prospective pilot starts by going to the local airport or to an aeronautical university to earn a *private pilot license.* A student pilot learns about aerodynamics, aviation regulations, airport operations, meteorology, radio communications, navigation, and of course, how to operate the airplane. A written exam, oral exam and flight test are required, and passing these allows the new pilot to rent or buy an airplane and take friends and relatives flying. The cost of instruction for a private pilot's license ranges from $3,500 to $5,000.

Commercial license

The next step is the *commercial pilot license.* There's not much to learn, but the pilot must demonstrate more experience and greater competency, since this license will allow him or her to fly for hire. For the flight test, the airplane used has retractable landing gear and flaps, as well as a propeller which the pilot controls separately from the engine. The pilot also takes, once again, a written and oral exam.

A commercial pilot license, by itself, is good for sight-seeing flights, towing banners or gliders, and basically local area flying in good weather.

Instrument rating

The *instrument rating* is the equivalent of a license that permits a pilot to fly in bad weather. It is required whenever a flight is conducted without reference to the ground. This occurs when flying in and above clouds, through rain, snow showers, fog, or other obstructions to visibility. Instrument flying means flying solely by reference to instruments in the cockpit and navigating by electronic guidance. Normally, if a person wants to fly professionally, he or she undergoes commercial and instrument training at the same time, at a cost of at least $6,000. The instrument rating requires—yes, once again—a written exam and oral exam, as well as a flight test. Either a private or commercial pilot may attempt to obtain it, and working on an instrument rating is an efficient way to gain the experience and competence necessary to achieve commercial certification.

Flight instructor's license

The next step up is the *flight instructor's license*. When applying for this license, the applicant needs to demonstrate the ability to *teach* how to fly an airplane. This involves doing commercial pilot maneuvers from the right seat rather than the pilot's typical left-side seat. And, once again, you must pass written, oral, and flight tests. This step costs about $2,000.

The new instructor pilot should now find a job instructing and building up flight time. (Flight time is a measure of experience, but does not always correlate directly with the flying ability of a pilot. Pilots with only 1,000 hours of flight time may fly more competently than others who have logged over 3,000 hours.) A brand new flight instructor can usually claim from 200 to 250 hours of flight time. He or she will need a minimum of 1,200 to 1,500 hours before anyone will consider him or her for employment in larger airplanes.

Multi-engine rating

At some point, our budding pilot will want to move up to a *multi-engine rating*. As the name implies, this license allows him or her to fly an airplane with more than one engine. The main focus of the training is on what to do if one engine fails. A multi-engine aircraft also has more systems and is therefore more complex than most single engine airplanes. The multi-engine airplane is usually faster and requires more of the pilot than most single engine airplanes. Getting to this step would cost about another $2,500 to $3,000.

Figure 19. **Cockpit** *The instruments in this cockpit of a Canadair Regional Jet use cathode ray tubes—like those found in television screens—to display information, rather than the needle indicators used for traditional instrumentation. Compare these instruments to those in* Figure 23 *(see also* **Question #46***).* Courtesy of Canadair

Instrument instructor license

The next step up is the *instrument instructor license,* an add-on to the flight instructor's license. This allows the instructor to teach students how to fly with reference to instruments only. An instrument instructor's students will be in training for their instrument rating.

Once he or she has accumulated enough time, the pilot may be hired to fly larger airplanes. My first airline job after instructing was flying a 15 passenger turboprop. It was a big change from what I had been doing, but it took me only a few hours of flying to get comfortable with it.

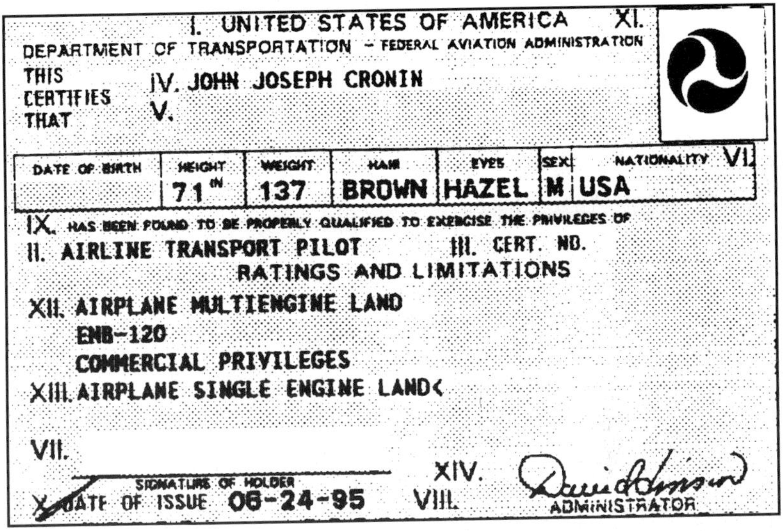
I. UNITED STATES OF AMERICA XI.
DEPARTMENT OF TRANSPORTATION - FEDERAL AVIATION ADMINISTRATION
THIS CERTIFIES THAT IV. JOHN JOSEPH CRONIN
V.

DATE OF BIRTH	HEIGHT	WEIGHT	HAIR	EYES	SEX	NATIONALITY VI.
	71 IN	137	BROWN	HAZEL	M	USA

IX. HAS BEEN FOUND TO BE PROPERLY QUALIFIED TO EXERCISE THE PRIVILEGES OF
II. AIRLINE TRANSPORT PILOT III. CERT. NO.
RATINGS AND LIMITATIONS
XII. AIRPLANE MULTIENGINE LAND
EMB-120
COMMERCIAL PRIVILEGES
XIII. AIRPLANE SINGLE ENGINE LAND<
VII.
SIGNATURE OF HOLDER
X. DATE OF ISSUE 06-24-95 VIII. XIV.
ADMINISTRATOR

Figure 20. ***Airline transport pilot license*** *The author's* airline transport pilot license, *the highest "degree" in commercial flight certification (see* Question #46*).*

Since then, I have steadily moved to a bigger, faster, and more complicated aircraft. The second airliner I flew was more modern than many jets and used TV screens instead of mechanical instruments (see *Figure 19*). Training for that aircraft and the jetliner I currently pilot is another whole story (see ***Question #47***).

Airline transport pilot license

The final step, and the ultimate in commercial flying licensing, is the Airline Transport Pilot License—it's like the Ph.D. of aviation (see *Figure 20*). This license is required to act as a captain at an airline or as the *pilot in command* of large aircraft and jets. It is earned on the basis of experience, a written test, and a flight test, usually in a multi-engine airplane.

Every license or rating involves flying with an official of the FAA (Federal Aviation Administration) or an FAA designated flight examiner (see ***Question #16***). A flight test is preceded by an oral

HYDRAULIC FAILURE

2. Loss Of The Green System

The loss of hydraulic systems caused by the loss of mechanical and electric hydraulic pumps or by hydraulic fluid leakage is indicated by:

MAIN PUMP LOW PRESS and ELEC PUMP lights or LOW LEVEL light illuminated on the hydraulic panel.

INOP light illuminated, below 120 KIAS, on the rudder panel.

HYD and RUDDER (below 120 KIAS) lights illuminated on the MAP.

CAUTION light flashing.

Green Hydraulic Pressure . . . . CHECK

If hydraulic pressure normal, refer to LOSS of HYDRAULIC MAIN PUMP

Loss of Green Hydraulic pressure means loss of:
- Landing gear retraction and normal extension
- Nosewheel steering
- Outboard normal brakes
- Outboard flaps
- Green rudder power
- Forward Door Actuator

Green System Electric Hydraulic Pump . . . . OFF

Approach and landing configurations:
    Gear; DOWN refer to ABNORMAL LANDING GEAR EXTENSION, FREEFALL.
    Flaps; UP
    Target Airspeed; 0 V_{REF} + 10 KIAS
    Reference Airspeed; 0 V_{REF}
GPWS 1 C/B (J25) . . . . PULL
Do not attempt to taxi

Figure 21. ***Systems failure checklist*** *Every aircraft comes equipped with a manual detailing a very specific protocol to follow in the event of an equipment or system failure (see* **Question #47***).*

exam which may last from one to two hours. With the exception of the multi-engine rating, all licenses and ratings require a written exam, and 80% or better is passing.

In all, making the transformation from landlubber to commercial airline pilot will cost you years of sweat and about $16,000.

47 What sort of training does an airline give a pilot?

All pilots must have either a commercial or airline transport pilot license when they are hired. Once hired, the pilot receives training in the type of airliner he or she will fly. Every airliner model is different. For example, a Boeing 747 pilot can't just hop into a DC-10 and fly away.

Pilots need to know exactly what they are doing when they turn a knob or flip a switch to activate a aircraft system, because it usually will have some effect on another system. Although pilots have specific checklists to follow when system failures occur (see *Figure 21*), it is critically important that they obtain a thorough knowledge of an aircraft's systems in case something goes wrong unexpectedly. This makes airline training essential.

Ground school
A new pilot, or one transferring from another aircraft type, will first attend *ground school* for two or three weeks. In ground school, pilots learn about each airliner system, for example, electrical and hydraulic systems. Pilots also learn about voltage, pressures, and temperatures, and when valves or electrical connections open and close. There may be several oral and written exams.

Ground school for the aircraft I currently fly lasted two weeks. I then spent about 28 hours in a cockpit procedures trainer, followed by more than a week of simulator flying. After that, I took an oral exam which lasted two hours. Then, I had a simulator session that mimicked an actual passenger flight. Finally, I spent several weeks flying passengers around with a *Check Airman Captain*. He or she is a captain certified to "break in" the pilots who will be flying an airliner which is new to them or who will be assuming a role that is new to them, i.e., a flight engineer who is becoming a copilot. Any failure means additional training is required. Two failures could result in termination of employment.

Cockpit procedures training

When the pilot has successfully completed ground school, he or she may attend *cockpit procedures training*, depending on the airline and aircraft. Cockpit procedures training makes use of a device which is just like the cockpit of the aircraft except nothing works (some lights and instruments may simulate actual operation). Its purpose is to aid the pilot in learning the location of every switch, button, knob, light, and instrument. Many switches have different shapes and sizes so the pilot can identify them by touch. The pilot learns whether activating a switch requires a pull, push, turn, or any combination of these maneuvers.

Simulator training

The final stage of training takes place in a fully operational simulator. From the outside, the simulator looks like a big, bumpy box on hydraulic lifts; inside, it is a fully realistic cockpit with a very realistic view out of the windows. The pilot learns to "fly" the airliner inside this simulator. He or she takes off, climbs to whatever altitude the instructor directs, and does some "airwork." This involves flying the airliner in a manner that would be prohibited if passengers were on board. One such maneuver is a steep turn, which may be performed as a complete turn in one direction with a bank of up to 60 degrees. This compares with a normal bank with passengers on board of about 15 to 25 degrees (see *Question #39*).

Once the pilot is comfortable "flying" the simulator under normal conditions, he or she learns to deal with failures and emergencies. When the simulator training is complete, the pilot undergoes an exam known as a *check-ride*. An examiner from the airline—and sometimes one from the Federal Aviation Administration—is present. The pilot takes an oral exam and then flies the simulator while the examiners observe.

Maintaining proficiency

It doesn't end there. Every six months, a captain must return to the simulator to remain proficient. A copilot does this every year. If either ever fail, they can conceivably be fired, but usually they'll get a second chance. Just before a pilot returns to the simulator, he or she attends a short ground school to remain current on airliner systems. Pilots also attend training to learn how to work as a team—this is known as "cockpit resource management."

Like most professions which deal with evolving technologies and responsibility for human life, a pilot's training never ends—and neither does the testing. If we fail even one simulator ride or medical exam, we can be summarily terminated. And if a pilot takes a job at another airline, he or she has to start at the bottom, regardless of experience. There is no such thing as job security for professional pilots.

48 How do pilots get their work schedule?

At the beginning of each month, airlines publish the possible schedules for pilots and copilots for the following month. Each potential schedule is called a *line* and there could be over 200 just for copilots of a particular airliner type (for example, a Boeing 727) at a particular base (for example, Atlanta). The copilot who is number one in seniority gets first choice. The folks at the bottom of the list get what's left. I happen to be number 139 out of 212. I make 139 choices—to make sure there's at least one schedule left over after the 138 pilots ahead of me have chosen—and hope to get one of my top 40 picks.

When a pilot reviews the schedules, he or she considers several factors, some of which are: the *report time*, the *end time*, number of days off, the specific days off, pay, amount of flying, and where

```
E7001  EXCPT FR SA SU          NOV 01-NOV 30              EXCEPT NOV 23 NOV 24
PILOT --> 0530      F/A --> 0530
1 727   330 EWR BOS 0630 0737  1:07  0:53
1 727   331 BOS EWR 0830 0954  1:24  0:41
1 727       EWR FLL 1035 1327  2:52         5:23  8:12  8:12      19:13  S1

PILOT --> 0810      F/A --> 0755
2 727  1165 FLL TPA 0840 0935  0:55  0:30
2 727      *TPA IAH 1005 1114  2:09  1:56
2 727   282 IAH CLE 1310 1642  2:32         5:36  8:47  9:02      14:08  S2
       SEE HOTEL LIST                       000-000-0000
PILOT --> 0620      F/A --> 0605
3 727   762 CLE EWR 0650 0814  1:24  0:51
3 727       EWR SRQ 0905 1203  2:58  0:42
3 727   763*SRQ EWR 1245 1529  2:44         7:06  9:24  9:39 SNK
                        TOTAL BLK:  18:05         TAFB:  58:14
```

Figure 22. **Pairings chart**   *Pilots get their schedules by choosing sets of flights called* pairings *from a detailed booklet which provides information on overnights, flight times, and departure and destination sites. Shown here is an excerpt of a* pairings chart *from one such booklet (see* **Question #48***).*

the overnights are. Pilots who commute to work are usually interested in a schedule that allows them to commute (fly) to their base the day they start work. (No one wants to commute on their day off and spend the night.) We also like to finish a workday in time to catch a flight back home. Some pilots want weekends off. Others want to maximize pay. As with any job, what every individual wants is different.

The pilots enter their *bids* into a computer, and then a few days later, the results are published. At this point, the pilot knows when he or she is off duty and in what cities he or she will be overnighting. The pilot also knows who their other crew members will be.

The final schedule is made up of *pairings* (see *Figure 22*). A pairing is a set of routes, or an itinerary, that takes one to four days to complete. Each pairing has a number on the schedule which can be used to reference the exact times and destinations for that pairing in a separate booklet. When the pilot sees that, for example, on November 14 he or she starts pairing number 7001, the pilot can look up that pairing in the booklet. When the pilot finds it, he or she sees every flight for which he or she has been scheduled. This guide shows departure date, time, and city; arrival date, time, and city; flight number; pay value; and hotel information.

49 **What do pilots carry in those square-like black cases?**

Mostly air charts and *approach plates*. Approach plates are navigational diagrams for a specific type of approach to a specific runway at a specific airport. In a big binder, I have many US airport approach plates. In a smaller binder, I carry the plates for airports I normally fly into, along with some alternates.

I also carry an aircraft operating manual, plus odds and ends such as a flashlight and spare eyeglasses. Additionally, I carry my own headset which is handy if the cockpit speaker should fail or if I have a problem with my microphone.

50 **How much control do pilots have over departure and arrival times?**

Very little. Just like the passengers we fly, we are at the mercy of air traffic control, the weather, passengers who arrive late at the gate, late bags, and late service from cleaners, catering, fueling, and lavatory personnel. If the airline instructs us to wait for some more bags, we wait. If air traffic control tells us we can expect to taxi thirty minutes late, so it goes.

The main thing we can control is making sure that we are not the cause of the delay. We can also try to make up time by asking air traffic control for more direct routing. For example, we might be willing to tolerate a little turbulence to get a better tailwind or reduce the effects of a headwind. But aside from these tactics, we are as much at the mercy of the scheduling gods as any passenger.

Figure 23. *Horizontal situation Indicator (HSI)* *Located at
the center of this photograph, the* horizontal situation indicator *is
one of the instruments used in conjunction with the* instrument
landing system (ILS), *which helps pilots align their aircraft with a
designated runway. In the very center of the HSI is an icon
representing the aircraft. As the airliner moves relative to the
runway, this aircraft icon shows the airliner's relationship to the
desired course. A pilot wants the* course deviation bar (**A**) *to
completely align with the* course pointer arrow (**B**). *The course
pointer will also line up exactly with the* heading marker (**C**)
providing no wind is present. The dotted vertical deviation pointer
and scale (**D**) *indicate displacement above or below the* glideslope
with a moving horizontal arrow (see **Questions** *#20 and #51).*

51 How does a pilot land in the fog?

A series of sophisticated electronic devices allow airliners to fly despite bad weather:

Instrument landing system

The *instrument landing system* (or *ILS*—see **Question #20**) has several components. Most notable are two antennae, one which gives course guidance to the runway, and another which gives guidance for descent to the landing end of the runway.

Upon arriving in the vicinity of the destination airport, the airliner is usually a few miles to the left or right of the course. The air traffic controller gives the pilot a heading in order to intercept and *join* the correct approach course.

Horizontal situation indicator

At this point, the pilot uses an instrument called an *omni course indicator* in smaller planes, or a *horizontal situation indicator* or *attitude director indicator* in larger planes, which receives and displays information from the ILS (see *Figure 23*). Using a series of overlapping lines, this device provides pilots with information on: 1) whether the airliner is aligned with, or is to the right or left of, the runway; and 2) whether the airliner is above, below, or directly on the *glideslope*, or path, to landing. As the plane approaches the course, a vertical line visible on this instrument begins to move towards the center of the instrument. When it reaches the instrument's center, the pilot is on course and lined up with the runway. The pilot now proceeds toward the runway, the pilot watching a horizontal line on top of his or her instrument. As he or she approaches the coordinates of precalculated descent information, the horizontal line begins to move down towards the center of the instrument. When it reaches the center, the pilot is again on the correct course for landing, the *glideslope* or *descent guidance* (see **Question #20**).

The instrument landing system usually allows the pilot to descend to within 200 feet above the ground. From there, the approach lights on the runway can usually be seen, allowing the pilot to land the airliner. If the pilot reaches this point and does not see the approach lights, then a missed approach must be executed (see *Question #43*).

Non-precision approaches

There are other approaches, referred to as *non-precision approaches*, which are used in poor weather, but none of these provide descent guidance. To execute a non-precision approach, the pilot still lines up with the runway, but descends to a predetermined altitude at a specific location, normally measured in miles from the navigation station in use. During the approach, the pilot descends to a final altitude and starts timing. When a predetermined time and distance have elapsed, the runway should be visible. If it isn't, a missed approach must be executed (see *Question #43*).

Glossary

aborted takeoff   the termination of a takeoff at any point after clearance for takeoff has been issued by *air traffic control*. It is executed any time a warning light appears, or any instrument or system begins to function abnormally.   *Question #41*

aileron   either of two small, flap-like devices located near the wingtips of an aircraft used to control rolling or *banking* movements.   *Question #40* and *Figure 15*

air cycle machine   device in which hot *bleed air* from an aircraft's engines travels through an expansion turbine, causing the air to become very cold. This cold air is combined with unprocessed hot bleed air in a mixing chamber to provide aircraft cabins with pressurized, room temperature air.   *Question #27*and *Figure 13*

air traffic control (ATC)   *air traffic control* is present at all major airports and at twenty-one additional radar facilities, nationally. Collectively, air traffic control monitors and regulates all air traffic departing and arriving in the United States, as well as all ground traffic within immediate airport boundaries.   *Question #12*

airline transport pilot license   the license required for a pilot to act as a captain for an airline, or as *pilot in command* of any large aircraft. This license is the final step in commercial flight certification.   *Question #46*

airspeed   aircraft speed in relation to the surrounding air, rather than to the ground.   *Question #11* and *Figure 3*

airway   a path, or continuous designated space of air through which aircraft are directed to fly by *air traffic control*. Called a *jetway* at altitudes above 18,000 feet.   *Question #2*

angle of attack   The angle between the relative wind and the wing, or the degree to which an aircraft's nose points up or down during flight.   *Question #37*

approach   procedures and/or instruments which allow an aircraft to arrive at the vicinity of a destination airport and to land there.   *Question #33*

approach control   the *air traffic control* personnel who are responsible for directing all incoming flights to the correct *approach path*, and for spacing incoming flights. *Question #12*

approach path   the actual course an aircraft takes as it descends for landing.   *Question #42*

approach plates   navigational diagrams which pilots use to reference the specific type of *approach* required for a specific runway at a specific airport.   *Question #49*

arrival push   airline traffic usually follows a pattern in which aircraft fly into and out of their airline's *hub* many times each day. The peak time of incoming traffic to a *hub* is called the *arrival push*.   *Question #17*

bank   a flight maneuver in which the pilot uses both of an aircraft's two *ailerons* to raise one wing and lower the other, thus causing the plane to turn. When an *aileron* is raised, its wing loses lift and moves down, turning the aircraft to that side. Simultaneously, the opposite *aileron* is lowered, increasing lift and raising its wing, which turns the aircraft to the opposite side.   *Question #39* and *Figure 17*

bleed air   hot air from a jet engine diverted away to *packs* and an *air cycle machine* where temperature is adjusted so that it can be used to provide climate control in an aircraft cabin. *Question #27* and *Figure 13*

center   any one of twenty-one *air traffic control* stations responsible for regulating all high altitude airspace in the United States.   *Question #12*

check-ride   the final exam a pilot takes at the end of *ground school* in which he or she operates a simulator or aircraft and takes an oral exam in the presence of an examiner from the *Federal Aviation Administration* or designee.   *Question #47*

cockpit procedures training (CPT)   a course of instruction for pilots which makes use of a simulated cockpit. Pilots use this equipment to learn the location, shape, and method of manipulation of every switch, button, and knob located in a specific aircraft's cockpit, as well as the appropriate sequence to follow when using them.   *Question #47*

commercial pilot license   flight certification that allows a pilot to fly for hire locally in good weather conditions. *Question #46*

cruise flight   the speed and altitude maintained by an aircraft for the majority of a flight.   *Question #24*

deferred maintenance item   an inoperative aircraft component which the *minimum equipment list* states must be repaired or replaced within a specified number of days or flights after failure.   *Question #26*

departure control   the *air traffic control* personnel responsible for monitoring all departing flights and for guiding departing aircraft to their initial *navigational fix*.   *Question #12*

departure push   airline traffic usually follows a pattern in which aircraft fly into, and out of, their airline's *hub* many times daily. The peak time of outgoing traffic is called the *departure push*.   *Question #17*

elevator   a moveable control surface mounted to the *horizontal stabilizer* (also known as the *tailplane*) which adjusts an aircraft's *pitch*.   *Question #39*and *Figure 15*

end time   fifteen minutes after the scheduled arrival time of the last flight of a pairing.   *Question #48*

Federal Aviation Administration (FAA)   the US government entity responsible for maintaining and regulating the common national civil-military system of air navigation and control.   *Question #16*

fix   see *navigational fix*.   *Question #12*

flap   the moveable control surfaces on the back of an aircraft's wing used for increasing *lift*, usually during takeoff and landing.   *Question #38*and *Figure 15*

flight instructor license   flight certification which allows a pilot to hire him- or herself out as an instructor. He or she must demonstrate competency in performing commercial pilot maneuvers from the cockpit's right seat (as opposed to pilot's traditional left seat).   *Question #46*

glideslope   the predetermined path an aircraft is intended to travel as it descends for landing. Also called *descent guidance*.   *Question #51*

go-around   an aborted landing attempt in which an aircraft ascends, then may circle to attempt another landing.   *Question #42*

ground controller the *air traffic control* personnel responsible for all movement on airport surfaces. *Question #12*

ground school the portion of pilot training which occurs in the classroom. *Question #47*

holding pattern a cyclical, oval-shaped path flown by an aircraft awaiting clearance to land, often during peak traffic hours and weather delays. *Question #14*

horizontal stabilizer This structure, attached to an aircraft's tail and shaped like an upside down wing, generates reverse *lift*, which pushes the rear of the aircraft towards the ground to keep it at the same level as the heavier front end of the aircraft. Also called a *tailplane*. *Question #39 and Figure 15*

horizontal situation indicator see *omni course indicator*. *Question #51 and Figure 23*

hub an airline's base of operations. An airline's *hub* is at an airport that houses a large number of its aircraft each night, and is the origin of a large number of that airline's connecting flights. *Question #17 and Figure 4*

instrument landing system (ILS) located at most major airports, this navigational equipment employs two separate antennae to provide pilots with vertical and horizontal guidance to the runway. *Questions #20 and Figure 8*

instrument rating flight certification which permits a pilot to fly without reference to the ground, such as during inclement weather or when visibility is poor. *Question #46*

instrument instructor license flight certification which allows a pilot to teach students to fly with reference to instruments only. *Question #46*

jetway   see *airway*.   *Question #2*

late check   designation given to any bag that is checked with an airline later than the airline's pre-flight deadline. *Late check* bags are not guaranteed to make their flights, and when they don't, are not delivered free of charge by most airlines. *Question #6*

leading edge slats and **flaps**   control surfaces which can be extended from the front of an aircraft's wings to increase lift during takeoff and landing.   *Question #37*

lift   the upward force generated by air passing over and under an aircraft's wings, resulting in air pressure above the wing decreasing in relation to that below. Lift is what allows airplanes to fly.   *Question #23; Figures 10 and 11*

local controller   *air traffic control* personnel located at any individual airport. These personnel are distinguished from those employed at any one of the twenty-one *air traffic control centers* in the United States.   *Question #12*

long haul market   that portion of the airliner and freight industry which flies long-distance domestic or international routes.   *Question #44*

minimum equipment list (MEL)   a comprehensive list generated for each airliner model, enumerating the parts or systems required for flight and under what conditions an aircraft may continue to fly when it malfunctions or fails.   *Question #26*

missed approach   an aborted landing attempt, executed after a pilot flying to a predetermined point of a final approach cannot see the runway.   *Question #43*

multi-engine rating   flight certification which allows the pilot to fly an aircraft with more than one engine. This training focuses on what actions to take in the event of an engine failure.   *Question #46*

National Transportation Safety Board   the premiere independent agency responsible for investigating all aircraft and major surface transport accidents.   *Question #16*

navigational fix   a specific location designated by latitudinal, longitudinal, and altitude information that is an intersection of two or more *airways* a specific distance from a navigation station. Airliners fly from their departure airport to such a location before continuing on to their final destination, passing through several other *fixes* on their way.   *Question #12*

omni course indicator   instrument used in conjunction with an *instrument landing system* in small aircraft. This device allows a pilot to align with the runway by showing an aircraft's position in relation to a horizontal and vertical line. In larger aircraft this information is given by other instruments called either the *horizontal situation indicator* or the *attitude director indicator*.   *Question #51*

outstation   all major airlines in the United States operate with hub and spoke route systems. *Hubs* are bases of operations, and the origins of connecting flights. The *spokes* represent connecting flights to *outstations* which are destinations that can only be reached from *hubs*.   *Question #17* and *Figure 4*

packs   the assembly into which hot *bleed air* from jet engines is diverted, and then mixed with *bleed air* that has been cooled in an *air cycle machine*. The mixed air is then directed into an airliner's passenger cabin.   *Question #27* and *Figure 13*

pairing   a pilot's schedule or a travel itinerary, representing a set of flights that takes one to four days to complete. Times, destinations, and layovers for each pairing can be referenced in a detailed book.   *Question #48* and *Figure 22*

pilot in command   the pilot ultimately responsible for all passengers, crew, and the aircraft itself. This person may actually be in the copilot's seat if the person in the captain's seat is in training to be a captain.   *Question #46*

private pilot license   flight certification which allows holder to rent, buy, and fly a private aircraft provided it is not used commercially.   *Question #46*

pitch   the degree to which an aircraft's nose points up or down as it pivots on its wings.   *Question #39*

propjet   see *turboprop*.   *Questions #25* and *#51*

prop brake   device which prevents a propeller from revolving while its engine is running.   *Question #31*

ramp   the large area of pavement located near an airport's gates.   *Question #12*

reduction gears   a series of interconnected different-sized gears that increase or decrease rotational speed between the first gear in the series and the last.   *Question #25*

report time   the time at which a flight crew must report for work, usually one hour before the first flight in a pairing and 90 minutes before all subsequent flights.   *Question #48*

reverse thrust   the force generated when a jet engine's exhaust is redirected toward an aircraft's nose, slowing the forward speed of the plane.   *Question #34*

rudder the moving part of an aircraft's tail which is attached to the *vertical stabilizer* (sometimes called the *tailfin*). The *rudder* moves either right or left, turning the body of the aircraft as it does so. *Question #40*and *Figure 15*

runway visual range (RVR) a measurement which informs pilots and *air traffic control* of the visibility level on the runway, calculated by a *transmissometer*. *Question #20*

seat mile a unit of measurement of airline usage representing one occupied seat in one airliner traveling one mile. An airliner with ten passengers which has traveled twenty miles therefore has accumulated 200 seat miles. *Question #44*

short haul market that portion of the airliner passenger and freight industry focusing on frequent short-distance flights. *Question #44*

slot one of several openings at the front of a wing surface that allows air from under the wing to flow upwards and over the wing's top surface. This allows aircraft to maintain lift even when flying with a high *angle of attack*, such as during takeoff and landing. *Question #37* and *Figure 16*

sorting area site to which all luggage for one airline is sent after being checked, to be placed on a cart for loading on to outgoing flights. *Question #5*

spoilers/speed brakes flat pieces of metal which can be raised from the wing to destroy, or *spoil* lift and create drag, slowing the airliner and/or helping the pilot move to a lower altitude. *Question #35*

static wick short wires protruding from an aircraft's wingtips and tail which discharge static electricity accumulated during flight. *Question #36* and *Figure 15*

tailfin   see *vertical stabilizer*.   *Question #39* and *Figure 15*

tailplane   see *horizontal stabilizer*.   *Question #39* and *Figure 15*

taxiway   paved airport surfaces which allow aircraft to travel between gate areas and runways.   *Question #19*

threshold   the beginning of the landing portion of a runway, marked with eight white stripes.   *Question #18* and *Figure 5*

through bags   bags which are identified in the *sorting area* as luggage needing to be directed to connecting flights. *Question #5*

thrust   the force generated when air is pushed rearward by jet engines or propellers, thus pushing an aircraft forward. When thrust is reversed, the aircraft is pushed backwards, which might only be perceived as a slowing of forward motion. *Questions #24 and #32*

timed-out   a mandatory eight-hour period of time when a flight crew rests between their last flight of the day and their first flight the next morning.   *Question #2*

tower control   the *air traffic control* personnel responsible for issuing takeoff and landing clearances, and for monitoring all air traffic within a five mile radius and up to an altitude of 2,500 feet.   *Question #12*

transmissometer   electronic device comprised of two platforms designed to measure a runway's visibility, or *runway visual range (RVR)*. One platform houses instruments which transmit light, while the other receives it. Based upon the amount of light received, this device calculates the RVR. *Question #20*

turboprop airliner   an airliner powered by a jet engine which turns a propeller to create thrust, also called a *propjet*. *Question #25*

turbulence   a disruption of smooth airflow, caused by rising and sinking columns of air and/or strong or gusty wind conditions. The result is a bumpy uncomfortable ride for passengers and crew.   *Question #10*

vertical stabilizer   the upright portion of an aircraft's tail which assists in maintaining direction (sometimes called a *tailfin*). The *rudder* is mounted to it.   *Question #39* and *Figure 15*

VHF omni-direction range (VOR)   navigation station which sends out signals based on magnetic compass headings to indicate whether pilots are on course. If they are not, the VOR supplies information for course correction. Located at some airports as well as in neighborhoods, empty fields and on mountains.   *Question #20* and *Figure 9*

vortex   a spiraling mass of gas or fluid which draws into its center all that surrounds it.   *Question #40*

vortex generators   small pieces of protruding metal mounted on an aircraft's wing and tail, designed to prevent separation of air from the wing and tail surfaces. This maintains optimal responsiveness of an aircraft's controls.   *Question #40* and *Figure 18*

wake turbulence   the tumultuous currents of air left behind an aircraft. *Wake turbulence* is a threat to all aircraft flying behind other aircraft. Small airliners must be most cautious while following other planes.   *Question #13*

weather radar   radar used specifically to evaluate weather conditions by measuring the amount and movement of precipitation in the air. *Question #8*

windshear a sudden shift in wind direction and/or velocity, which can cause unexpected changes in an aircraft's airspeed and altitude during takeoff and landing. *Question #11* and *Figures 2* and *3*

Index

About the Author

John Cronin is a graduate of Daniel Webster College with a degree in aviation management and flight training. He has earned private, commercial, instrument, multi-engine, and airline transport pilot licenses. His first job as an airline pilot was flying the Beech 99C from Cape Cod to other New England destinations. He has also piloted Embraer's EMB-120 Brasilia and has worked as a flight engineer on a Boeing 727. He now copilots Boeing 737s for a major US airline.

John lives in the New York City area with his wife Terry; their two daughters, Kelly and Courtney; and their dog, Taffy.

Combat Aircraft Recognition
3rd edition
by Peter March

The definitive guide to the recognition of military aircraft flown around the world by nations great and small. They're all here, from the McDonnell Douglas F-15 to the Boeing B-52 to the MiG-29 and European Tornado fighters, and more. *Includes choppers as well as fixed wing craft.*

4 1/2" by 7", 108 pages, soft cover
ISBN 1-882663-26-8
Item #PP-CAR3 .... **$14.95**

Classic & Warbird Recognition
by Peter March

A great guide to take to airshows and air museums, or just to browse at home. This book covers vintage aircraft—both civilian and military—which are kept in flying condition by individuals and museums around the world. Each entry offers details of the history and performance capabilities of these beloved planes. *Includes over 100 photos of the classic birds in flight.*

4 1/5" by 7", 96 pages, soft cover
ISBN 0-7110-2423-5
Item #IA-CLWB **$12.95**

Light Aircraft Recognition
3rd edition
by Peter March

This guide to recognition gives all the information that spotters will need to identify the small prop planes and helicopters commonly seen on airport runways...and in local skies. *Includes such well known craft as the Piper Cherokee, Piper Super Cub, Aeronca Champion, and the Cessna 150 and 180.*

4 1/2" by 7", 108 pages, soft cover
ISBN 1-882663-15-2
Item #PP-LAR3 .... **$11.95**

Airliner Note Cards
Featuring beautiful color paintings by Richard King

We commissioned our favorite aviation artist to create six new airliner paintings especially for these note card sets. Modern airliners make up one set, while beloved vintage airliners are depicted in the other. Cards are of heavy stock, measure 5 by 6 5/8 inches folded, and are packed in attractive, sturdy boxes. *Each set includes 9 note cards and 9 envelopes, 3 each of 3 paintings.*

Jetliners

Vintage Airliners

Modern Jetliners set *includes Boeing 747, Airbus A300, McDonnell Douglas DC-9*
ISBN 1-882663-18-7
PP-JET **$9.95**

Vintage Airliners set *includes Douglas DC-3, Ford Trimotor, China Clipper*
ISBN 1-882663-19-5
PP-VINT **$9.95**

United 747-400
DA-UA747 .... **$19.95**

Airliner Models!
from Wooster and Aviation International

These beautiful scale airliner models are finished in exacting detail with authentic airline liveries. Each comes with an attractive display stand, making your airliner model ready for flight on any desk or shelf. Collect a whole fleet from important airlines such as American, United, Northwest, Delta, TWA, Continental, and US Airways, including such famous and beloved craft as the Boeing 747, 757, and 767; the Douglas DC-3, DC-9, and DC-10; the McDonnell-Douglas MD-11 and MD-80; and the Airbus A300, A320, and A340. *Model prices vary.*

- **Over 110 different models to choose from**
- **More than 45 different airline liveries**

abc Airline Series!

Each book in this series spotlights one of the world's major airlines. With words and pictures—many in color—the reader is transported from the humble origins of each of the airlines through its evolution into a large and important carrier. *Included for each carrier:* information on current operations, fleet configurations, route systems, and more.

Series specs: 7 1/4" by 4 3/4", 96 pages, soft cover

abc American Airlines
by Simon Forty
ISBN 1-882663-21-7
Item #PP-AA $12.95

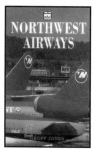

abc Northwest Airlines
by Geoff Jones
ISBN 1-882663-28-4
Item #PP-NA $12.95

abc United Airlines
by Simon Forty
ISBN 1-882663-20-9
Item #PP-UA $12.95

abc Delta Airlines
by Geoff Jones
ISBN 1-882663-29-2
Item #PP-DA $12.95

abc British Airways
by Leo Marriott
ISBN 0-7110-2510-X
Item #IA-BA $12.95

Famous Airliners, 2nd edition

by William F. Mellberg

Bill Mellberg turns his story-telling skills to chronicling the evolution of the modern airliner. **Famous Airliners** begins with the Boeing Model 80, whose 12 passengers flew at a little over 100 mph and were the first to enjoy the attention of specially trained nurses called "stewardesses." It ends with the Concorde, which travels at more than twice the speed of sound and over the Atlantic in about three hours. Covering more than 70 airliners, this book includes over 150 color and black and white photographs of vintage and modern aircraft. Printed on high contrast gloss paper, the beautiful second edition now includes such important classic craft as the Zeppelin Hindenburg and Curtiss Condor, as well as modern jetliners such as the Boeing models 757, 767, and 777.

Praise for the first edition:

"...will appeal to youngsters and aviation novices (as well as to) airline enthusiasts and frequent flyers." Airways magazine

"Tells the story of airliner development with perspective and accuracy..." Robert Bradford, Retired Director, National Aviation Museum, Canada

7" by 10", 240 pages, soft cover, ISBN 1-882663-13-6
Item #PP-FA2 .... **$24.95**

Boeing Airliners

747/757/767 in Color

by Alan J. Wright & Robbie Shaw

Boeing is a name that is synonymous with the giant jetliners which traverse the world's airways. The company's Boeing 747, with its humped forward fuselage, has become an icon of the jet age. This book covers the continuously updated 747; and its modern siblings, the narrow-bodied 757 and wide-bodied 767. Includes over 100 color pictures.

6 3/4" by 9 1/4", 128 pages, soft cover, ISBN 1-882663-24-1
Item #PP-BAL **$24.95**

More great aviation books and gifts from Plymouth Press...

Guide to Airport Airplanes
by William & Frank Berk

Includes color photos and 3-view silhouettes of commonly observed airliners

Facilitating identification of airliners with its simple-to-use, systematic approach, the *Guide* features the 66 most commonly observed airliners, all pictured in color photographs while in flight or at interesting airport locales. Capabilities, as well as country of origin and date of first flight are included, making the *Guide* an excellent basic reference as well as an indispensable airport companion.

"A great aid in aircraft identification."
Hemispheres (United Airline's in-flight magazine)

"A handy guide that can be taken along on trips for quick identification." Toronto Star

"A marvelous reference book..." Robert W. Bradford, retired Director, National Aviation Museum, Canada

7" by 5", 168 pages, soft cover, ISBN 1-882663-10-1
Item #PP-GAP2 .... **$14.95**

The Airport Airplane Coloring Book
by Richard King

Travelers and airliner buffs can bring the romance of the airport home or have their airport experience enhanced by *The Airport Airplane Coloring Book*. It includes airliners often observed at major airports, such as the Boeing 727, 737, and 747; the McDonnell-Douglas DC-9 and DC-10; and the Airbus A300.

"Hours of fun coloring...the perfect gift for an aviation minded youngster." Airliners magazine

8 1/2" by 11", 44 pages, soft cover, ISBN 1-882663-05-5
Item #PP-AACC **$5.95**